Series in BioEngineering

The Series in Bioengineering serves as an information source for a professional audience in science and technology as well as for advanced students. It covers all applications of the physical sciences and technology to medicine and the life sciences. Its scope ranges from bioengineering, biomedical and clinical engineering to biophysics, biomechanics, biomaterials, and bioinformatics.

Indexed by WTI Frankfurt eG, zbMATH.

More information about this series at https://www.springer.com/series/10358

Yongjin Zhou · Yong-Ping Zheng

Sonomyography

Dynamic and Functional Assessment of Muscle Using Ultrasound Imaging

Yongjin Zhou
Health Science Center
Shenzhen University
Shenzhen, China

Yong-Ping Zheng
Department of Biomedical Engineering
Research Institute for Smart Ageing
Hong Kong Polytechnic University
Kowloon, Hong Kong

ISSN 2196-8861 ISSN 2196-887X (electronic)
Series in BioEngineering
ISBN 978-981-16-7142-5 ISBN 978-981-16-7140-1 (eBook)
https://doi.org/10.1007/978-981-16-7140-1

This Springer imprint is published by the registered company Springer Nature Singapore Pte Ltd.
The registered company address is: 152 Beach Road, #21-01/04 Gateway East, Singapore 189721, Singapore

Preface

It is well known in musculoskeletal clinical assessment and research that muscle functions are closely related to their structural features. Modern medical imaging modalities can reveal various structural information in the human body, including that of muscle. In sports medicine and rehabilitation science, ultrasound imaging (USI) is a commonly used technique to visualize skeletal muscles under different conditions as it is noninvasive, real time, easy to use, and less costly compared to other imaging modalities for muscle, such as MRI. However, musculoskeletal studies using USI used to be mainly qualitative; i.e., subjective observation of the static ultrasound frames by experienced experts is the common practice for the diagnosis or investigations of skeletal muscle.

As ultrasound images are nowadays almost all in digital forms, the rapid development of digital image processing, especially in the field of computer vision, has promoted the research progress in the studies of skeletal muscle using USI. In 2006, Zheng and his team demonstrated that morphological changes of forearm muscles from ultrasound images can be used as indicators of activity for both able-bodied individuals and amputees [1]. The report demonstrated a successful application of cross-correlation technique for the automatic dynamic tracking of the thickness change in skeletal muscle during contraction. Firstly, it is an objective and quantitative study using USI. Secondly, its computation is conducted in real time which is beneficial and helpful in subsequent applications of motion intent recognition and prosthetic control. The term sonomyography (SMG) was proposed as a counterpart to the more well-known technique for functional assessment of muscle, namely electromyography (EMG). EMG is an electrodiagnostic technique for recording and evaluating the electrical activity produced by skeletal muscle during contraction, while SMG signal reflects the structural changes of skeletal muscle detected using USI in real time. EMG and SMG can provide complementary information about muscle functions. One unique advantage of SMG is its ability to resolve activities of neighboring or overlapping muscles, which is difficult to be achieved by EMG due to the inherent combination of EMG signals generated by muscles which are overlapped or close to each other.

Since then, Zheng's team and other research groups in the field have begun to work on the development of novel techniques for quantifying the static as well as dynamic structural information of skeletal muscles associated with different contraction tasks. Specifically, the dimension of the SMG has developed from 1D and 2D to 3D. Besides EMG, other signals including muscle force/torque and inertial sensing signals can also be collected simultaneously with SMG. The image processing involved also advanced from traditional computer vision methods to more recent techniques aided by deep learning and artificial intelligence to improve measurement accuracy as well as computational efficiency. In addition, the research work has been extended to clinical applications from initially tracking algorithm development.

Over the years, through substantial collaborations with collaborators working in many different medical and healthcare fields, the huge gap has been gradually bridged by the technique developers and healthcare professionals, paving the way for a wider application of sonomyography techniques.

It would have been impossible for the authors to complete this book without all the research and development work conducted by students: Jun Shi, Mathew Chan, Congzhi Wang, Jingyi Guo, Guangquan Zhou, Qinghua Huang, Zengtong Chen, Chenglang Yuan, Connie Cheng, Zihao Huang, Xiaocheng Yu, Jane Ling, Queenie Shea, Shuai Li, Longjun Ren, and many others. A group of postdoctoral fellows have also contributed to the research work, including Drs. Xin Chen, Haris Begovic, Tianjie Li, Hongbo Xie, Christina Ma, and together with a group of engineers, including Like Wang, Junfeng He, Zhengming Huang, etc. They have worked directly on SMG development and applications over the years since 2006, and many of them have become professors and made great achievements in their research fields. Chenglang Yuan also directly contributed to Chap. 3, Zengtong Chen to Chaps. 2 and 4, and Xiaocheng Yu and Shiyu Sun to Chap. 7. We also thank Yang De and Lyn Wong for their supports in editing of symbols and managing copyright matters of figures, respectively.

The authors would also like to express their sincere thanks to all their collaborators in different fields, and all these collaborations inspired them to continue improving techniques for quantitative ultrasound in skeletal muscle, which also led to the writing of this book. Finally, we deeply appreciate the continuous supports, encouragements, and patience from our families over the years to our academic career development!

Shenzhen, China   Yongjin Zhou
Hong Kong, China   Yong-Ping Zheng

Reference

[1]   Y.-P. Zheng, M. Chan, J. Shi, X. Chen, and Q.-H. Huang, "Sonomyography: Monitoring morphological changes of forearm muscles in actions with the feasibility for the control of powered prosthesis," *Medical engineering & physics,* vol. 28, no. 5, pp. 405–415, 2006.

Contents

Abbreviations

2D NCC	Normalized two-dimensional cross-correlation
AT	Achilles tendon
AUC	Area under the curve of ROC
BM	Brachialis muscle
BMI	Body mass index
B-mode	Brightness mode
CMC	Coefficient of multiple correlations
C-MI-FFD	Constrained mutual-information-based free-form deformation
CNN	Convolutional neural network
CNR	Contrast-to-noise ratio
CSA	Cross-sectional area
CT	Computerized tomography
DCNN	Deconvolutions and max-unpooling
DSC	Dice similarity coefficient
EF	Extra features
EFOV US	Extended field-of-view ultrasonography
EMG	Electromyography
FL	Fascicle length
FOS	First-order statistic
GLCM	Gray-level co-occurrence matrix
GLNU	Gray-level non-uniformity
GLRLM	Gray-level run length matrix
GM	Gastrocnemius muscle
HM	Hamstring muscle
HT	Hough transform
ICC	Intra-class correlation coefficient
IPAQ	International Physical Activity Questionnaire
LBP	Local binary pattern
L_f	Fascicle length
LGIF	Local and global intensity fitting
LOO	Leave-one-out

LRE	Long-run emphasis
LRT	Localized radon transform
MC	Megacycle
MFAF	Mean frequency analysis feature of the ROI
MG	Medial gastrocnemius
MI-FFD	Mutual information-based free-form deformation
MIFS	Multiple images feature selection
MMG	Mechanomyography
MRI	Magnetic resonance imaging
MT	Muscle thickness
MUI	Musculoskeletal ultrasound image
MVC	Maximum voluntary contraction
MVEF	Multiscale vessel enhancement filtering
NCC	Normalized cross-correlation
NIH	National Institutes of Health
PA	Pennation angle
pdfs	Probability density functions
peak-to-average	Peak amplitude and the average amplitude
QM	Quadriceps muscle
QMT	Quadriceps muscle thickness
ResNet	Residual convolutional network
RF	Radiofrequency
RFE	Recursive feature elimination
RFM	Rectus femoris muscle
RLNU	Run length non-uniformity
ROC	Receiver operating characteristic
ROI	Region of interest
RP	Run percentage
RT	Radom transform
RVHT	Revoting Hough transform
SAR	Synthetic-aperture radar
SD	Standard deviation
SEM	Standard error of the measurement
SHT	Standard Hough transform
SM	Soleus muscle
SMG	Sonomyography
SMMG	Sonomechanomyography
SNR	Signal–noise ratio
SOL	Soleus
SPECT	Radioisotope imaging
SRE	Short-run emphasis
SSI	Supersonic shear imaging
SVM	Support vector machine
TA	Tibialis anterior
TAF	Texture analysis features of the ROI

TC	Tetanic contraction
US	Ultrasound
USI	Ultrasound imaging
VI	Vastus intermedius
VIM	Vastus intermedius muscle
VL	Vastus lateralis
VM	Vastus medialis
Xcorr	Cross-correlation
X-CT	X-ray computed tomography

Chapter 1
Automatic and Quantitative Methods for Sonomyography (SMG)

Abstract The history and recent development of Sonomyography (SMG) are introduced in this chapter. The increasing demands arising from the progress of musculoskeletal research has interacted with the rapid development of contemporary ultrasound imaging, computer vision, and AI technologies over the last few decades. This interaction has propelled a series of successful applications of medical ultrasound, which has further promoted musculoskeletal research. All these together have formed a very interesting and dynamic research field.

Ultrasound imaging (USI) has been widely used in fundamental studies on healthy skeletal muscles, as well as examinations of pathology or trauma in muscles. Medical imaging techniques such as X-ray computerized tomography (CT) [1, 2] and magnetic resonance imaging (MRI) [3–5] have also been used to obtain the parameters of musculotendon complex in vivo. However, compared with CT and MRI, USI technology is radiation-free, real-time, low-cost, versatile in operation circumstances, and widely available and relatively low-cost; therefore, it has been a more popular method in the measurement of muscle morphological or structural parameters [6]. Meanwhile, the validity of ultrasound imaging measurement has been verified in comparison with direct anatomical measurements using cadavers [7–9].

Propelled by the advancements in image processing techniques and imaging equipment themselves, there have appeared recently some interesting trends in skeletal muscle studies using ultrasound imaging. The first is the increasing usage of automatic methods in the estimation of pennation angle (PA), muscle thickness (MT), fascial length (FL) and cross-sectional area (CSA) of skeletal muscles, linking with various musculoskeletal conditions. The second is about the data nature, advancing from individual slices to image sequences captured. The third is conditions of the studied muscles, extending from healthy ones, whose features are relatively more known or well modeled, to ones in pathological or abnormal statuses.

In this review, we present a summary of recent studies in the morphological information of skeletal muscles using ultrasound imaging, mainly focusing on the achievements of the automatic estimation of morphological parameters. A preliminary report is also presented on how these achievements can possibly bridge morphological information to the functional activities during muscle contractions and how

© Springer Nature Singapore Pte Ltd. 2021
Y. Zhou and Y.-P. Zheng, *Sonomyography*, Series in BioEngineering,
https://doi.org/10.1007/978-981-16-7140-1_1

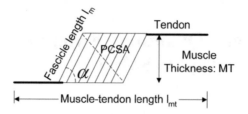

Fig. 1.1 Simplified structure of a musculotendon actuator. I_{mt}: muscle–tendon length; I_m: muscle fiber length; α: pennation angle; PCSA: physiological cross-section area. Republished with the permission of EUREKA SCIENCE (FZC), from N. Chen, H. Hu, and L. Li, "Ultrasound Imaging of Muscle–tendon Architecture in Neurological Disease: Theoretical Basis and Clinical Applications," Current Medical Imaging, vol. 10, no. 4, pp. 246–251, 2014; permission conveyed through Copyright Clearance Center, Inc

the recent advancements in engineering fields have facilitated SMG up to a level of potentially triggering changes in medical practices.

USI has been used to measure changes in many morphological parameters which include muscle PA and fascicle orientation, MT, FL, fascicle curvature [10, 11], cross-sectional area [12] and muscle size [13, 14] or volume, as illustrated in Fig. 1.1.

Fascicle orientation or PA is among the first group of studied parameters using automatic methods. Traditionally, the lines and angles in musculoskeletal ultrasound images were detected manually [15, 16], or interactively using software [17, 18], such as NIH Image (National Institutes of Health, Bethesda, MD, USA; http://rsb.info.nih.gov/nih-image). Using the software, the orientation of a manually drawn line on the studied image could be read. These methods are time-consuming and the manual detecting process is subjective by nature. Thus, automatic estimation methods for musculoskeletal morphology are desired for further development and application [19].

1.1 Popular Skeletal Muscles Studied Using Ultrasound Imaging

USI has been applied to studies of various muscles including the gastrocnemius muscle (GM), quadriceps muscle (QM), rectus femoris (RF), brachialis muscle (BM), vastus lateralis (VL), Achilles tendon (AT), etc. The GM has gained the most popularity among various skeletal muscles. Table 1.1 describes various muscles studied with USI and 72 research works in total have been identified from Web of Science with the topics of 'muscle' and 'architecture' and titles including the word 'ultrasound' from 2004 to 2019.

Table 1.1 The number of different research projects on various skeletal muscles

Muscle	Amount of research	Details
Gastrocnemius muscle (GM)	30	[10–13, 20–45]
Quadriceps muscle (QM)	10	[36, 37, 46–53]
Rectus femoris (RF)	6	[13, 28, 37, 45, 47, 54]
Brachialis muscle (BM)	3	[27, 40, 55]
Vastus lateralis (VL)	5	[26, 30, 45, 48, 53]
Soleus muscle (SM)	3	[12, 29, 39]
Hamstring muscle (HM)	1	[7]
Tibialis anterior (TA)	6	[26, 37, 43, 45, 56, 57]
Achilles tendon (AT)	2	[20, 25]

1.2 Estimation of Muscle Architecture and Contraction

Though most reports assume the muscle fascicles in ultrasound to be straight (for example, [58–60]), some mathematical models [61] and experimental studies [62–65] have indicated that fascicles may in fact have curves. Therefore, Darby and colleagues [66] developed an automated approach without strong shape constraints to permit accurate characterization of dynamic changes in geometric properties of muscle. A study on the medial gastrocnemius, a commonly studied muscle in the lower limb, and a Bayesian tracking framework was used to quantify fascicle behavior in vivo during a wide range of movements.

Meanwhile to better model muscle motion, Rana et al. tried to compute fascicle orientation in 3D space, using a 2D ultrasound scanner and a 3D position tracker [30, 31]. The 3D parameter can be achieved with short scan times (less than 2 min for the gastrocnemii) and would thus enable future studies to quantify 3D muscle architecture during sub-maximal voluntary contractions.

Meanwhile, Zhou et al. tried to extend the studies on muscle architectural parameters toward its functional counterpart [67, 68] by studying the architectural dynamics during muscle contractions. The remaining research question aims to understand how the architecture dynamics affect muscle function, which is challenging but meanwhile significant for clinical applications.

1.3 For Muscles of Aged People or Subjects with Pathological Muscle Conditions

It is a direct benefit of the advancements in the computerized processing methods on ultrasound image sequence that continuous PA, MT and FL signals similar to electromyography (EMG), mechanomyography (MMG) or torque can be acquired. The sampling interval equals the frame rate of the ultrasound sequence. In other words, it becomes possible now to tag and analyze the muscle contraction dynamics with ultrasound too, besides the traditional EMG, MMG, or torque signals. Moreover, the former includes valuable spatial information with muscle specificity. This merit is also one of the reasons why ultrasound is growing in popularity in both sports science and rehabilitation. However, a major limitation of current automatic methods is that most achievements in the automatic computation of morphological parameters are achieved on young and healthy subjects (for example, [14, 27, 29, 33–35, 69, 70]), whose ultrasound images exhibit strong and relatively stable pennate appearance. The patterns can be dramatically undermined by ageing or pathological conditions. Solutions for a wider range of images remain both desirable and challenging.

Only a few studies have made preliminary efforts to automatically estimate muscle parameters for muscles from aged people or post-stroke or subjects with other pathological conditions [18, 37, 68, 71].

In Chaps. 2, 3, 4, 5, 6, 7 and 8, we introduce the development of SMG in several important aspects. Specifically, muscle thickness, pennation angle, fascicle length, cross-sectional area, enhancement of skeletal ultrasound images, some exploratory AI applications in SMG, and finally future perspectives of SMG are discussed.

References

1. Narici, M.V., Maganaris, C.N., Reeves, N.D., Capodaglio, P.: Effect of aging on human muscle architecture. J. Appl. Physiol. **95**(6), 2229–2234 (2003)
2. Sunnerhagen, K.S., Svantesson, U., Lönn, L., Krotkiewski, M., Grimby, G.: Upper motor neuron lesions: their effect on muscle performance and appearance in stroke patients with minor motor impairment. Arch. Phys. Med. Rehabil. **80**(2), 155–161 (1999)
3. Kawakami, Y., Nakazawa, K., Fujimoto, T., Nozaki, D., Miyashita, M., Fukunaga, T.: Specific tension of elbow flexor and extensor muscles based on magnetic resonance imaging. Eur. J. Appl. Physiol. **68**(2), 139–147 (1994)
4. Maganaris, C.N., Baltzopoulos, V., Sargeant, A.J.: Changes in the tibialis anterior tendon moment arm from rest to maximum isometric dorsiflexion: in vivo observations in man. Clin. Biomech. **14**(9), 661–666 (1999)
5. Graichen, H., Englmeier, K., Reiser, M., Eckstein, F.: An in vivo technique for determining 3D muscular moment arms in different joint positions and during muscular activation—application to the supraspinatus. Clin. Biomech. **16**(5), 389–394 (2001)
6. Narici, M.V., Binzoni, T., Hiltbrand, E., Fasel, J., Terrier, F., Cerretelli, P.: In vivo human gastrocnemius architecture with changing joint angle at rest and during graded isometric contraction. J. Physiol. **496**(1), 287–297 (1996)
7. Kellis, E., Galanis, N., Natsis, K., Kapetanos, G.: Validity of architectural properties of the hamstring muscles: correlation of ultrasound findings with cadaveric dissection. J. Biomech. **42**(15), 2549–2554 (2009)
8. Ema, R., et al.: In vivo measurement of human rectus femoris architecture by ultrasonography: validity and applicability. Clin. Physiol. Funct. Imaging **33**(4), 267–273 (2013)
9. Ando, R., Taniguchi, K., Saito, A., Fujimiya, M., Katayose, M., Akima, H.: Validity of fascicle length estimation in the vastus lateralis and vastus intermedius using ultrasonography. J. Electromyogr. Kinesiol. **24**(2), 214–220 (2014)
10. Bénard, M.R., Becher, J.G., Harlaar, J., Huijing, P.A., Jaspers, R.T.: Anatomical information is needed in ultrasound imaging of muscle to avoid potentially substantial errors in measurement

of muscle geometry. Muscle Nerve Official J. Am. Assoc. Electrodiagnostic Med. **39**(5), 652–665 (2009)

11. Namburete, A.I., Rana, M., Wakeling, J.M.: Computational methods for quantifying in vivo muscle fascicle curvature from ultrasound images. J. Biomech. **44**(14), 2538–2543 (2011)

12. De Oliveira, L.F., Menegaldo, L.L.: Individual-specific muscle maximum force estimation using ultrasound for ankle joint torque prediction using an EMG-driven Hill-type model. J. Biomech. **43**(14), 2816–2821 (2010)

13. Rosenberg, J.G., Ryan, E.D., Sobolewski, E.J., Scharville, M.J., Thompson, B.J., King, G.E.: Reliability of panoramic ultrasound imaging to simultaneously examine muscle size and quality of the medial gastrocnemius. Muscle Nerve **49**(5), 736–740 (2014)

14. English, C., Fisher, L., Thoirs, K.: Reliability of real-time ultrasound for measuring skeletal muscle size in human limbs in vivo: a systematic review. Clin. Rehabil. **26**(10), 934–944 (2012)

15. Chen, N., Hu, H., Li, L.: Ultrasound imaging of muscle-tendon architecture in neurological disease: theoretical basis and clinical applications. Curr. Med. Imaging **10**(4), 246–251 (2014)

16. Fukunaga, T., Kawakami, Y., Kuno, S., Funato, K., Fukashiro, S.: Muscle architecture and function in humans. J. Biomech. **30**(5), 457–463 (1997)

17. Manal, K., Roberts, D.P., Buchanan, T.S.: Can pennation angles be predicted from EMGs for the primary ankle plantar and dorsiflexors during isometric contractions? J. Biomech. **41**(11), 2492–2497 (2008)

18. Itoi, E., Sashi, R., Minagawa, H., Shimizu, T., Wakabayashi, I., Sato, K.: Position of immobilization after dislocation of the glenohumeral joint: a study with use of magnetic resonance imaging. JBJS **83**(5), 661–667 (2001)

19. Reeves, N.D., Narici, M.V., Maganaris, C.N.: In vivo human muscle structure and function: adaptations to resistance training in old age. Exp. Physiol. **89**(6), 675–689 (2004)

20. Zhou, Y., Zheng, Y.-P.: Estimation of muscle fiber orientation in ultrasound images using revoting hough transform (RVHT). Ultrasound Med. Biol. **34**(9), 1474–1481 (2008)

21. Bolsterlee, B., Veeger, H.D., van der Helm, F.C., Gandevia, S.C., Herbert, R.D.: Comparison of measurements of medial gastrocnemius architectural parameters from ultrasound and diffusion tensor images. J. Biomech. **48**(6), 1133–1140 (2015)

22. Chino, K., Akagi, R., Dohi, M., Takahashi, H.: Measurement of muscle architecture concurrently with muscle hardness using ultrasound strain elastography. Acta Radiol. **55**(7), 833–839 (2014)

23. Crisóstomo, R., Candeias, M., Armada-da-Siva, P.: The use of ultrasound in the evaluation of the efficacy of calf muscle pump function in primary chronic venous disease. Phlebology **29**(4), 247–256 (2014)

24. Gillett, J.G., Barrett, R.S., Lichtwark, G.A.: Reliability and accuracy of an automated tracking algorithm to measure controlled passive and active muscle fascicle length changes from ultrasound. Comput. Methods Biomech. Biomed. Eng. **16**(6), 678–687 (2013)

25. Huijing, P.A., Bénard, M.R., Harlaar, J., Jaspers, R.T., Becher, J.G.: Movement within foot and ankle joint in children with spastic cerebral palsy: a 3-dimensional ultrasound analysis of medial gastrocnemius length with correction for effects of foot deformation. BMC Musculoskelet. Disord. **14**(1), 1–12 (2013)

26. Kwah, L.K., Pinto, R.Z., Diong, J., Herbert, R.D.: Reliability and validity of ultrasound measurements of muscle fascicle length and pennation in humans: a systematic review. J. Appl. Physiol. **114**(6), 761–769 (2013)

27. Li, L., Tong, K.Y., Hu, X.: The effect of poststroke impairments on brachialis muscle architecture as measured by ultrasound. Arch. Phys. Med. Rehabil. **88**(2), 243–250 (2007)

28. Li, Q., et al.: Continuous thickness measurement of rectus femoris muscle in ultrasound image sequences: a completely automated approach. Biomed. Sig. Process. Control **8**(6), 792–798 (2013)

29. Peixinho, C.C., Martins, N.S.F., de Oliveira, L.F., Machado, J.C.: Structural adaptations of rat lateral gastrocnemius muscle–tendon complex to a chronic stretching program and their quantification based on ultrasound biomicroscopy and optical microscopic images. Clin. Biomech. **29**(1), 57–62 (2014)

30. Rana, M., Hamarneh, G., Wakeling, J.M.: Automated tracking of muscle fascicle orientation in B-mode ultrasound images. J. Biomech. **42**(13), 2068–2073 (2009)
31. Rana, M., Wakeling, J.M.: In-vivo determination of 3D muscle architecture of human muscle using free hand ultrasound. J. Biomech. **44**(11), 2129–2135 (2011)
32. Yang, Y.-B., Zhang, J., Leng, Z.-P., Chen, X., Song, W.-Q.: Evaluation of spasticity after stroke by using ultrasound to measure the muscle architecture parameters: a clinical study. Int. J. Clin. Exp. Med. **7**(9), 2712 (2014)
33. Yu, J.-Y., Jeong, J.-G., Lee, B.-H.: Evaluation of muscle damage using ultrasound imaging. J. Phys. Ther. Sci. **27**(2), 531–534 (2015)
34. Zhou, G.-Q., Chan, P., Zheng, Y.-P.: Automatic measurement of pennation angle and fascicle length of gastrocnemius muscles using real-time ultrasound imaging. Ultrasonics **57**, 72–83 (2015)
35. Zhou, Y., Li, J.-Z., Zhou, G., Zheng, Y.-P.: Dynamic measurement of pennation angle of gastrocnemius muscles during contractions based on ultrasound imaging. Biomed. Eng. Online **11**(1), 1–10 (2012)
36. Cho, K.H., Lee, H.J., Lee, W.H.: Reliability of rehabilitative ultrasound imaging for the medial gastrocnemius muscle in poststroke patients. Clin. Physiol. Funct. Imaging **34**(1), 26–31 (2014)
37. Kim, M.K., Ko, Y.J., Lee, H.J., Ha, H.G., Lee, W.H.: Ultrasound imaging for age-related differences of lower extremity muscle architecture. Phys. Therapy Rehabil. Sci. **4**(1), 38–43 (2015)
38. Peixinho, C.C., da Fonseca Martins, N.S., de Oliveira, L.F., Machado, J.C.: Reliability of measurements of rat lateral gastrocnemius architectural parameters obtained from ultrasound biomicroscopic images. Plos one **9**(2), e87691
39. Peixinho, C., Ribeiro, M., Resende, C., Werneck-de-Castro, J., De Oliveira, L., Machado, J.: Ultrasound biomicroscopy for biomechanical characterization of healthy and injured triceps surae of rats. J. Exp. Biol. **214**(22), 3880–3886 (2011)
40. Li, L., Tong, K., Hu, X., Hung, L., Koo, T.: Incorporating ultrasound-measured musculotendon parameters to subject-specific EMG-driven model to simulate voluntary elbow flexion for persons after stroke. Clin. Biomech. **24**(1), 101–109 (2009)
41. Antonios, T., Adds, P.J.: The medial and lateral bellies of gastrocnemius: a cadaveric and ultrasound investigation. Clin. Anat. **21**(1), 66–74 (2008)
42. Barber, L., Barrett, R., Lichtwark, G.: Validation of a freehand 3D ultrasound system for morphological measures of the medial gastrocnemius muscle. J. Biomech. **42**(9), 1313–1319 (2009)
43. Botter, A., Vieira, T.M.M., Loram, I.D., Merletti, R., Hodson-Tole, E.F.: A novel system of electrodes transparent to ultrasound for simultaneous detection of myoelectric activity and B-mode ultrasound images of skeletal muscles. J. Appl. Physiol. **115**(8), 1203–1214 (2013)
44. Fry, N., Gough, M., Shortland, A.: Three-dimensional realisation of muscle morphology and architecture using ultrasound. Gait Posture **20**(2), 177–182 (2004)
45. Molinari, F., Caresio, C., Acharya, U.R., Mookiah, M.R.K., Minetto, M.A.: Advances in quantitative muscle ultrasonography using texture analysis of ultrasound images. Ultrasound Med. Biol. **41**(9), 2520–2532 (2015)
46. Debernard, L., Robert, L., Charleux, F., Bensamoun, S.F.: Characterization of muscle architecture in children and adults using magnetic resonance elastography and ultrasound techniques. J. Biomech. **44**(3), 397–401 (2011)
47. Delaney, S., Worsley, P., Warner, M., Taylor, M., Stokes, M.: Assessing contractile ability of the quadriceps muscle using ultrasound imaging. Muscle Nerve **42**(4), 530–538 (2010)
48. e Lima, K.M., Carneiro, S.P., Alves, D.d.S., Peixinho, C.C., de Oliveira, L.F.: Assessment of muscle architecture of the biceps femoris and vastus lateralis by ultrasound after a chronic stretching program. Clin. J. Sport Med. **25**(1), 55–60 (2015)
49. Benjafield, A., Killingback, A., Robertson, C., Adds, P.: An investigation into the architecture of the vastus medialis oblique muscle in athletic and sedentary individuals: an in vivo ultrasound study. Clin. Anat. **28**(2), 262–268 (2015)

50. Connolly, B., et al.: Ultrasound for the assessment of peripheral skeletal muscle architecture in critical illness: a systematic review. Crit. Care Med. **43**(4), 897–905 (2015)
51. Infantolino, B.W., Challis, J.H.: Estimating the volume of the First Dorsal Interossoeus using ultrasound. Med. Eng. Phys. **33**(3), 391–394 (2011)
52. Engelina, S., Antonios, T., Robertson, C., Killingback, A., Adds, P.: Ultrasound investigation of vastus medialis oblique muscle architecture: an in vivo study. Clin. Anat. **27**(7), 1076–1084 (2014)
53. Infantolino, B.W., Gales, D.J., Winter, S.L., Challis, J.H.: The validity of ultrasound estimation of muscle volumes. J. Appl. Biomech. **23**(3), 213–217 (2007)
54. Chi-Fishman, G., Hicks, J.E., Cintas, H.M., Sonies, B.C., Gerber, L.H.: Ultrasound imaging distinguishes between normal and weak muscle. Arch. Phys. Med. Rehabil. **85**(6), 980–986 (2004)
55. Li, L., Tong, K.: Musculotendon parameters estimation by ultrasound measurement and geometric modeling: application on brachialis muscle. In: 2005 IEEE Engineering in Medicine and Biology 27th Annual Conference, pp. 4974–4977. IEEE
56. Klimstra, M., Dowling, J., Durkin, J.L., MacDonald, M.: The effect of ultrasound probe orientation on muscle architecture measurement. J. Electromyogr. Kinesiol. **17**(4), 504–514 (2007)
57. Liu, P., Wang, Y., Mao, Y.: The muscle architectural parameters and strength changes of tibialis anterior in stroke survivors as measured by ultrasound. Chin. J. Rehabil. Med. 8
58. Blazevich, A.J., Gill, N.D., Zhou, S.: Intra-and intermuscular variation in human quadriceps femoris architecture assessed in vivo. J. Anat. **209**(3), 289–310 (2006)
59. Loram, I.D., Maganaris, C.N., Lakie, M.: Use of ultrasound to make noninvasive in vivo measurement of continuous changes in human muscle contractile length. J. Appl. Physiol. **100**(4), 1311–1323 (2006)
60. Zhou, G.-Q., Zheng, Y.-P.: Automatic fascicle length estimation on muscle ultrasound images with an orientation-sensitive segmentation. IEEE Trans. Biomed. Eng. **62**(12), 2828–2836 (2015)
61. Van Leeuwen, J., Spoor, C.: Modelling mechanically stable muscle architectures. Philos. Trans. Royal Soc. London Ser. B Biol. Sci. **336**(1277), 275–292
62. Kawakami, Y., Ichinose, Y., Fukunaga, T.: Architectural and functional features of human triceps surae muscles during contraction. J. Appl. Physiol. **85**(2), 398–404 (1998)
63. Muramatsu, T., Muraoka, T., Kawakami, Y., Shibayama, A., Fukunaga, T.: In vivo determination of fascicle curvature in contracting human skeletal muscles. J. Appl. Physiol. **92**(1), 129–134 (2002)
64. Wang, H.-K., Wu, Y.-K., Lin, K.-H., Shiang, T.-Y.: Noninvasive analysis of fascicle curvature and mechanical hardness in calf muscle during contraction and relaxation. Man. Ther. **14**(3), 264–269 (2009)
65. Stark, H., Schilling, N.: A novel method of studying fascicle architecture in relaxed and contracted muscles. J. Biomech. **43**(15), 2897–2903 (2010)
66. Darby, J., Li, B., Costen, N., Loram, I., Hodson-Tole, E.: Estimating skeletal muscle fascicle curvature from B-mode ultrasound image sequences. IEEE Trans. Biomed. Eng. **60**(7), 1935–1945 (2013)
67. Li, J., Zhou, Y., Zheng, Y.-P., Li, G.: An attempt to bridge muscle architecture dynamics and its instantaneous rate of force development using ultrasonography. Ultrasonics **61**, 71–78 (2015)
68. Li, J., Zhou, Y., Ivanov, K., Zheng, Y.-P.: Estimation and visualization of longitudinal muscle motion using ultrasonography: a feasibility study. Ultrasonics **54**(3), 779–788 (2014)
69. Zhao, H., Zhang, L.-Q.: Automatic tracking of muscle fascicles in ultrasound images using localized radon transform. IEEE Trans. Biomed. Eng. **58**(7), 2094–2101 (2011)
70. Zhou, G., Zheng, Y.-P.: Human motion analysis with ultrasound and sonomyography. In: 2012 Annual International Conference of the IEEE Engineering in Medicine and Biology Society. IEEE, pp. 6479–6482

71. English, C.K., Thoirs, K.A., Fisher, L., McLennan, H., Bernhardt, J.: Ultrasound is a reliable measure of muscle thickness in acute stroke patients, for some, but not all anatomical sites: a study of the intra-rater reliability of muscle thickness measures in acute stroke patients. Ultrasound Med. Biol. **38**(3), 368–376 (2012)

Chapter 2
Measurement of Skeletal Muscle Thickness

Abstract The change of muscle thickness (MT) can effectively reflect muscle activity during contraction, and is an important index for musculoskeletal studies in sports science or rehabilitation engineering. This chapter introduces the development of MT measurement technology from multiple perspectives, including: 1D to 2D and cross-sectional to longitudinal plane of muscles. The transverse cross-sectional perspective introduces the application of cross-correlation technology, compression tracking, and other technologies. The longitudinal perspective introduces the methods based on the revoting Hough transform, method based on MI-FFD registration, and local and global intensity fitting (LGIF) method, etc. These technologies developed in the field of computer vision have been successfully applied in musculoskeletal ultrasound and strongly promote the automated process of muscle thickness measurement.

2.1 Manual Measurement and Applications

Muscle thickness (MT), which can effectively reflect muscle activity during muscle contraction [1–4], is an important measure for musculoskeletal studies using ultrasonography. There are reports on the relationships between MT and many other measures, such as level of physical activity [5], age [6], weight [7], gender [7], muscle stiffness [8], muscle strength [7, 9], and response to exercise [10]. Muscle thickness is an important determinant of muscle condition in many areas. For example, Miyatani et al. used muscle thickness to estimate the muscle volume of the quadriceps femoris based on ultrasound imaging (USI) [11, 12]. Shi et al. used muscle thickness detected by ultrasound images to characterize the behavior of muscles when they are under fatigue [13]. Ohata et al. employed muscle thickness to quantify the muscle strength of people with severe cerebral palsy [14]. English et al. validated that ultrasound is a reliable measure of muscle thickness in acute stroke patients for some anatomical sites [15]. Freilich et al. found statistically significant relationships between quadriceps MVC (the isometric maximum voluntary contraction), quadriceps thickness, and body weight [7].

© Springer Nature Singapore Pte Ltd. 2021
Y. Zhou and Y.-P. Zheng, *Sonomyography*, Series in BioEngineering,
https://doi.org/10.1007/978-981-16-7140-1_2

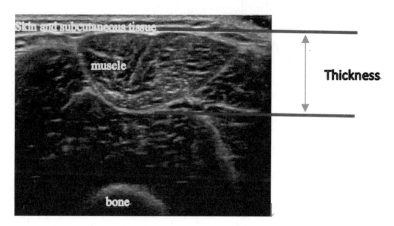

Fig. 2.1 Ultrasonography image showing various structures and measurement of muscle thickness

USI using a real-time brightness mode (B-mode) has a similar capability to computerized tomography (CT) scanning or magnetic resonance imaging (MRI) in visualizing fat and muscle tissues, with the advantage that USI can measure in real time. Muscle thickness, defined as the distance between two fascias, is easily determined with USI [9]. Let's take the manual measurement of quadriceps muscle thickness as an example. A circumferential mark was put at the midway between the tip of the greater trochanter and the lateral joint line of the knee. A linear probe was placed on this circumferential line, perpendicular to the skin and moved along the line till a suitable image was obtained. Then, the point corresponding to the center of the probe was marked with a vertical line. This point was used as the reference for all subsequent measurements of quadriceps thickness. The skeletal muscle thickness of the quadriceps muscle between the superficial fat–muscle interface and the femur was measured anteriorly [16], as shown in Fig. 2.1.

A real-time B-mode image depicts a cross-section of anatomical structures. In contrast, M-mode uses a narrow beam to produce a one-dimensional view of anatomical structures over time. M-mode can be used simultaneously with B-mode to enable the operator to accurately adjust the position of the M-mode beam, which is indicated by a dotted line on the B-mode display (Fig. 2.2). Information from sound waves reflected back to the transducer from anatomical structures in the US beam is used to produce a depth versus time chart of these structures in the M-mode display. The movement of structures in M-mode is one-dimensional and provides information on structural movement towards or away from the transducer. Orientating the plane of the US beam to capture the movement of anatomical structures of interest enables information to be obtained, over time, in a single image [17]. We can use the M-mode image to measure dynamic muscle thickness changes manually by selecting a number of critical points in the M-mode image for the interfaces between subcutaneous fat and muscle as well as between muscle and bone. Then the two groups of manually selected points can be used for fitting the interfaces so as to obtain the change of

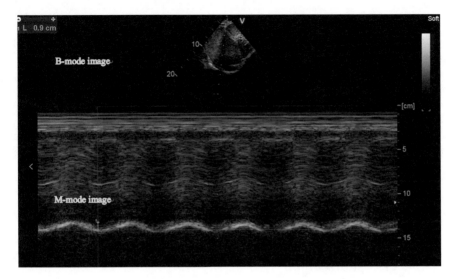

Fig. 2.2 An M-mode image simultaneously displayed with a B-mode image

the thickness by measuring the distance of two fitted curves in the depth direction. Manual drawing of contours in M-mode images is time-consuming and difficult. Hence, echocardiographers usually identify the wall boundaries only at end-systole and end-diastole for heart assessment [18].

Ultrasonography has been successfully used to measure the thickness of human skeletal muscle in vivo [19, 20]. However, it requires the observer to select the measurement site accurately, and to find the interfaces between subcutaneous fat and muscle as well as between muscle and bone [21], which makes manual measurement of muscle thickness subjective and time-consuming.

Some studies show that USI is a reliable tool for the measurement of quadriceps muscle thickness by critical care physicians with excellent inter- and intra-class reliability [16], but others have pointed out that the result of manually measuring muscle thickness by ultrasound can be reliably assessed by the same observer, but not between observers [22].

Previous studies have demonstrated that the muscle thickness change detected by USI during contraction, namely sonomyography (SMG), can be used for functional assessment of skeletal muscles and has the potential for prosthetic control [23-25]. SMG signals of skeletal muscle have approximate linear relationships with the corresponding joint angle [23]. A simple linear algorithm can be used to map the magnitude of SMG signal to the prosthesis opening position. Therefore, the relationship between the control signal, namely the wrist angle and the target signal of prosthesis position is approximately linear. This intuitive control method helps subjects to perform better with fewer conscious efforts. To satisfy similar requirements in EMG control, much more complex algorithms such as finite state machine should be applied and more training efforts are required [26]. Compared with EMG control, the major advantage of SMG control is that it is more suitable for providing intuitive control, because it is directly extracted from the morphologic changes of muscle, which are inherently closely correlated to the motion of joints. Intuitiveness relieves the mental burden on a user during long-term operation and natural daily work. Another advantage of SMG control is the capability of detecting contraction of individual muscles at neighboring locations and different depths, providing the feasibility of fine control in multifunction artificial hands. Due to the challenges in EMG signals generated by different neighboring muscles, i.e., cross talk, commercially available prostheses controlled by EMG can only provide a limited number of degrees of freedom [27].

2.2   1D Sonomyography

2.2.1   Cross-Correlation

Cross-correlation (Xcorr) measures the similarity of two signals over time. It is an important analytical tool in time-series signal processing as it can highlight when two signals are correlated with each other but exhibit some delay from one another (Fig. 2.3). The cross-correlation between two signals $u(t)$ and $v(t)$ is calculated by:

$$w(t) = u(t) \otimes v(t) \triangleq \int_{-\infty}^{\infty} u^*(\tau)v(\tau + t)d\tau \tag{2.1}$$

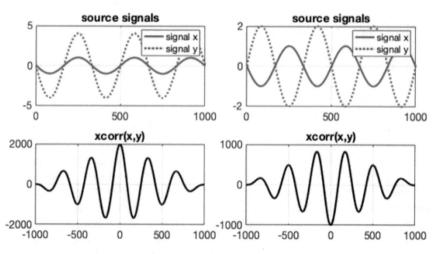

Fig. 2.3   The cross-correlation between two signals $u(t)$ and $v(t)$

The complex conjugate, $u*(\tau)$ makes no difference if $u(t)$ is real-valued but makes the definition work even if $u(t)$ is complex-valued. Cross-correlation is similar in nature to the convolution of two functions. In an autocorrelation, which is the cross-correlation of a signal with itself, there is always a peak at a lag $(w(t),$ delay of u versus $v)$ of zero, and its magnitude represents the energy of signal.

2.2.2 Echo Tracking

In 1972, Hokanson et al. presented a phase-locking device that tracked a particular zero crossing within the radiofrequency (RF) signals [28]. This technique improved the radial resolution, down to a few micrometers. Echo tracking involves serial measurements of the location (depth) of a particular echo structure, obtained from periodic pulsing in a single direction. Measurement of changes in aponeuroses involves tracking the distance between two ultrasound echo structures (one for each fascia). The resolution of the method depends on the temporal resolution of the receiver, rather than the frequency and bandwidth of the ultrasound pulse. Therefore, although conventional 5–10 MHz ultrasonic imaging systems have an axial resolution of about 0.3–0.1 mm, echo tracking boasts a resolution in the order of 0.001 mm for measuring changes in echo position, i.e. thickness change. The high resolution of echo tracking explains its popularity for measurement of contour [29].

The cross-correlation methods estimate the displacement by locating the maximum of the cross-correlation function between two successive radiofrequency signals. In general, the location of the maximal cross-correlation does not coincide with the sampling grid and the maximum has to be found by interpolating the sampled cross-correlation function. By detecting small displacements of the fascia at different times with cross-correlation methods, the aponeuroses can be tracked to measure accurately the dynamic change of muscle thickness.

2.2.3 M-mode

M-mode imaging involves the same pulsing sequence as echo tracking, but the processing of returning echoes is significantly different. M-mode imaging discards the phase information of the returning echoes and displays the envelope of the echoes as image intensity [29].

Semiautomatic contour detection has the potential to simplify the work of extracting continuous wall boundaries. Contour detection is often modeled as an energy minimization problem. Unser and colleagues described a method based on template matching [30]. The correlation between a reference template and the pixel values in the neighborhood of a candidate point gave a local energy value. The sum of all local energy values along a candidate contour defined a global energy, and the

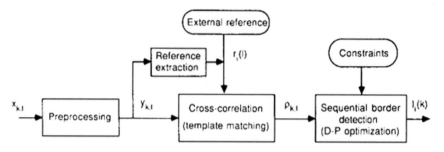

Fig. 2.4 Block diagram of a general system for sequential border detection for M-mode imaging. ©
2021 IEEE. Reprinted, with permission, from M. Unser, G. Pelle, P. Brun, and M. Eden, "Automated
extraction of serial myocardial borders from M-mode echocardiograms," IEEE Transactions on
Medical Imaging, vol. 8, no. 1, pp. 96–103, 1989

optimal contour was found by dynamic programming as the candidate contour that
had the lowest global energy value (Figs. 2.4 and 2.5) [31].

2.3   2D Sonomyography

2.3.1   *Transverse Cross-Sectional Imaging*

Transverse cross-sectional images of muscles can be easily obtained using ultrasound, by placing the ultrasound probe perpendicular to the long axis of the muscle
on the muscle belly, as shown in Fig. 2.6. By a coarse-to-fine method based on
a compressive-tracking algorithm for estimation of muscle thickness (MT) changes
[32], we can detect the MT changes in a cross-sectional plane using ultrasonography.
Figure 2.6 shows an ultrasound image of the quadriceps muscle as an example.

    As a sensitive and efficient solution for the detection of QM thickness changes in
the cross-sectional plane, the coarse-to-fine method for the automatic detection of MT
changes includes two stages: coarse locating and fine-tuning, as shown in Fig. 2.7.
The first stage of the method is to extract coarse location using the compressive-
tracking algorithm [33]. Here, the tracking problem is formulated as a detection
task. In the first frame of each ultrasound image sequence, three initial tracking
windows are selected manually along the top and bottom of the RF muscle and
femur, respectively. In this selection procedure, each window is placed such that its
horizontal center line lies at the exact boundary of the RF muscle or femur, as shown
in Fig. 2.8. The sizes of these windows are chosen empirically to include sufficient
features for reliable tracking. At each incoming frame, some positive samples (near
the current target location) and negative samples (far away from the object center)
are sampled to update the classifier. To predict the object location in the next frame,
some samples around the current target location are drawn and the one with the
maximal classification score is then treated as the expected location.

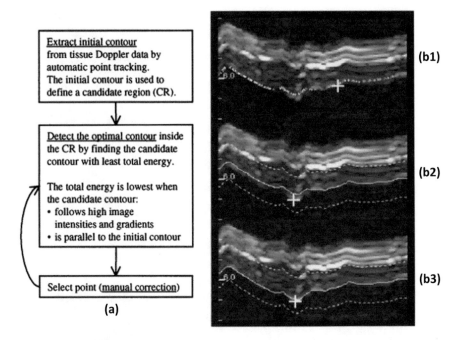

(a)

Fig. 2.5 **a** Flow chart of the semiautomatic contour-detection method. **b1** The operator moves the cursor (+) to a boundary in the image. The cursor is the starting point for extraction of the initial contour (...). (**b2–b3**) Automatic detected result, (....) the candidate region; (—) the detected optimal contour; (+) a point that the operator selects for manual correction. The optimal detected contour is forced through this point (lower right). Reprinted from S. I. Rabben et al., "Semiautomatic contour detection in ultrasound M-mode images," Ultrasound in Medicine and Biology, vol. 26, no. 2, pp. 287–296, 2000, with permission from Elsevier

Fig. 2.6 An example of a cross-sectional image of the quadriceps muscle

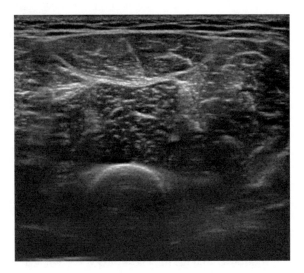

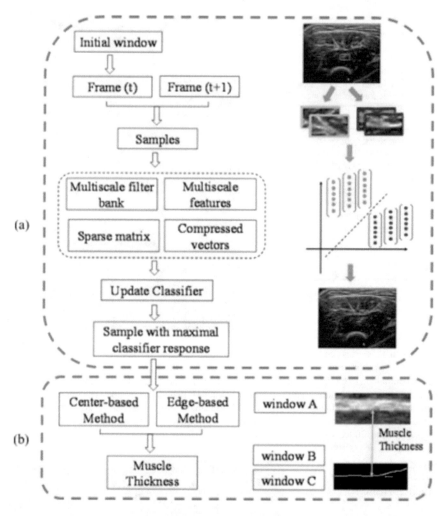

Fig. 2.7 Diagrammatic illustrations of the muscle thickness measurement method using a course-to-fine approach. **a** Coarse position-locating of the tracked window using a compressive-tracking algorithm, and **b** Fine-tuning to estimate the MT using the center-based method and edge-based method. © 2021 IEEE. Reprinted, with permission, from J. Li, Y. Zhou, Y. Lu, G. Zhou, L. Wang, and Y.-P. Zheng, "The sensitive and efficient detection of quadriceps muscle thickness changes in cross-sectional plane using ultrasonography: a feasibility investigation," IEEE Journal of Biomedical and Health Informatics, vol. 18, no. 2, pp. 628–635, 2013

The MT at each moment is defined as the vertical distance among window locations. Specifically, the thickness of the rectus femoris (RF) is defined as the maximum vertical distance between window A and window B, the thickness of the vastus intermedius muscle (VIT) is defined as the maximum vertical distance between window B and window C, and the thickness of QM is defined as the minimum vertical distance

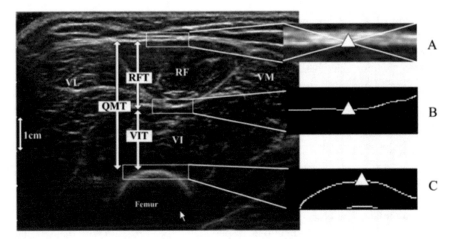

Fig. 2.8 Illustration of MT definitions. Three initial tracking windows A, B, and C are selected to obtain rectus femoris muscle (RFT), vastus intermedius muscle (VIT), and quadriceps muscle thickness (QMT). B and C have been handled by the Canny edge detector and the maximal connected components search technology. The triangles point out the locations of the selected window A, window B, and window C, respectively. © 2021 IEEE. Reprinted, with permission, from J. Li, Y. Zhou, Y. Lu, G. Zhou, L. Wang, and Y.-P. Zheng, "The sensitive and efficient detection of quadriceps muscle thickness changes in cross-sectional plane using ultrasonography: a feasibility investigation," IEEE Journal of Biomedical and Health Informatics, vol. 18, no. 2, pp. 628–635, 2013

between window A and window C. The locations of these windows are determined in two different ways, which are stated as follows and illustrated in Fig. 2.7:

1. Edge-based method: The compressive-tracking algorithm provides a rough location of the tracked window. Then, for windows B and C, the Canny edge detector and maximal connected components search are used to extract the edge. The Canny edge detector transfers the window image to a binary image, and then maximal connected components search technology is used to find the exact boundary of interest, as illustrated in Fig. 2.8.
2. Center-based method: There exist multiple boundaries for window A, after applying the Canny edge detector. Therefore, the location of window A is calculated as the center of this corresponding window, as illustrated in Fig. 2.8.

2.3.2 Normalized Cross-Correlation (NCC)

A cross-correlation algorithm can be used to track the displacements of the tissue regions of interest in the images. It has been reported that the correlation tracking algorithm can work for images with low signal noise ratio (SNR) and with complex structures of target and background [34], and is particularly useful for ultrasound images with complex speckles [23]. It requires a reference image or template (containing the object of interest) from an initial image frame and looks for the most similar area to the reference image in order to estimate the object's position in the subsequent frames. The equation used to calculate the normalized two-dimensional cross-correlation (2D NCC) is as follows [13]:

$$R(i, j) = \frac{\sum_{m=0}^{M-1} \sum_{n=0}^{N-1} [x(m, n) - \overline{X}][y(m, n) - \overline{Y}]}{\sqrt{\sum_{m=0}^{M-1} \sum_{n=0}^{N-1} [x(m, n) - \overline{X}]^2 \sum_{m=0}^{M-1} \sum_{n=0}^{N-1} [y(m + i, n + j) - \overline{Y}]^2}}$$

(2.2)

where $x(m, n)$ and $y(m + i, n + j)$ $(m = 0, M,$ and $n = 0, N)$ are the pixels of the selected image blocks in two different frames. The image block $x(m, n)$ is regarded as the template. The block represented by $y(m + i, n + j)$ shifts by i and j pixels in the horizontal and vertical directions, respectively, in comparison with the template represented by x(m, n). $\overline{X}$ and $\overline{Y}$ represent the means of pixel density for the image blocks $x(m, n)$ and $y(m + i, n + j)$, respectively, while $R(i, j)$ is the cross-correlation coefficient between them. By changing i and j, the correlation coefficients between the template in the first image and a group of image blocks in the second image can be calculated. The best matched image block of the template can then be located according to the peak value of the correlation coefficients.

To track muscle thickness changes, two vertically aligned rectangular windows, one along the superior and one along the inferior boundary of a muscle, were selected manually at the first image. The superior and inferior windows were placed such that their horizontal center lines lay at the superior and inferior hyperechoic muscle aponeuroses, respectively, and there was a good contrast between the muscle boundary and other surrounding structures [35], and a cross-correlation algorithm was used to measure the distance between their centers as muscle thickness at each moment (Fig. 2.9).

However, for sensitive and efficient detection of MT changes, the NCC method has several drawbacks. Apart from an intensive calculation requirement, a more general problem with NCC is the choice of the tracking window size [36]. The larger the window size, the more robust while less sensitive to fine changes of MT the tracking is, and vice versa. It is believed that both sensitivity to and efficiency of fine changes detection of MT are crucial in muscle study based on ultrasound image sequences, which are commonly acquired nowadays at a frame rate of 12–30 Hz and the interframe muscle motion is usually relatively small.

2.3.3 Sparse Tracking

Tracking can be considered as an estimation of the state for a time series state space model. The problem is formulated in probabilistic terms. Early works used a Kalman filter to provide solutions that are optimal for a linear Gaussian model. The

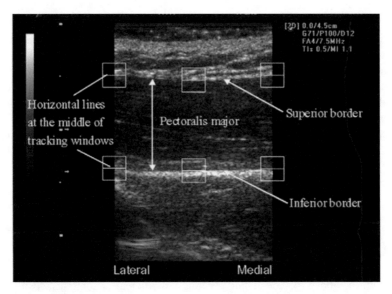

Fig. 2.9 An ultrasound image with boundary tracking windows superimposed on it. Each window has a horizontal line at its center that defines the boundary level. Thickness at each aspect is the distance between the two horizontal lines. Reprinted from T. K. Koo, C. Wong, and Y. Zheng, "Reliability of sonomyography for pectoralis major thickness measurement," Journal of Manipulative and Physiological Therapeutics, vol. 33, no. 5, pp. 386–394, 2010, with permission from Elsevier

particle filter, also known as the sequential Monte Carlo method [33], is one of the most popular approaches. It recursively constructs the posterior probability density function of the state space using Monte Carlo integration. It has been developed in the computer vision community and applied to tracking problems under the name Condensation [37]. In a previous study, an appearance-adaptive model was incorporated in a particle filter to realize robust visual tracking and classification algorithms [38].

Tracking can also be treated as finding the minimum distance from the object to be tracked to the subspace represented by the training data or previous tracking results [39, 40]. Rose et al. presented a tracking method that incrementally learned a low-dimensional subspace representation, efficiently adapting in real time to changes in the appearance of the target [41]. It is common that a muscle may change shape and appearance in ultrasound images as the contraction level changes, thus tracking methods with adaptive capability have certain advantages. However, adaptive tracking may potentially accumulate errors, as the improper adaptive adjustment in each step can add up.

Sparse representation has been successfully applied in numerous vision applications [42–45]. With sparsity constraints, one signal can be represented in the form of linear combination of only a few basis vectors. In previous studies [16, 17], the target candidate was sparsely represented as a linear combination of the atoms of a dictionary which was composed of dynamic target templates and trivial templates. This sparse representation problem was then solved through $\ell 1$ minimization with non-negativity constraints. In another study [46], dynamic group sparsity which includes both spatial and temporal adjacency was introduced into the sparse representation to enhance the robustness of the tracker. Sparse representations are effective models to account for appearance changes, and local sparse representations are more effective than the ones with holistic sparse templates [47].

Compared with other methods, tracking using sparse representation can more robustly and rapidly track muscle aponeuroses in ultrasound images, thereby calculating the muscle thickness more stably.

2.4   Longitudinal Imaging

2.4.1   Hough Transform

Although the above methods using tracking are efficient in some tasks, still many cumulative tracking errors arise owing to uneven muscle deformation, speckle noise interference, and even significant changes in the appearance or intensity of targets among subsequent images. Moreover, because of the differences in the operator's experience, the semi-automatic interaction while manually selecting the target tracking point in the first frame might be biased.

During the measurement of muscle aponeurosis, the detection of line structures is frequently involved, which is a fundamental problem in computer vision. The Hough transform (HT) [48] of slope–intercept parameterization and its improved version, HT of angle–radius parameterization [49, 50], have become an established solution to the problem.

Standard Hough transform uses the normal parameterization of a straight line in an image [49],

$$x \cos \theta_1 + y \sin \theta_1 = \rho_1 \tag{2.3}$$

where θ_1 is the angle between the normal of the line and the x-axis, and ρ_1 is the distance of the origin of coordinates to the line. The parameters of all straight lines going through a point (x_i, y_i) in the image space show up in the (ρ, θ) parameter space as a sinusoidal curve, given by:

$$x_i \cos \theta + y_i \sin \theta = \rho \tag{2.4}$$

After transforming all edge/feature points to the (ρ, θ) space, the collinear points will cross each other and an array measuring the crossing situation is accumulated. Traditionally this array (ρ, θ) is called an accumulator array. The next stage of the SHT is an exhaustive search for the maxima in the accumulator array, and a predefined threshold is set so that all local values of $H(\rho, \theta)$ exceeding the threshold can be recognized as evidence of straight lines existing in the original image space.

In summary, in SHT, the collinear edge/feature points in the image space show up as peaks in the (ρ, θ) space. However, in the realization of SHT, there are some issues: (i) digital image is discrete by nature; (ii) when mapped into (ρ, θ) space, ρ also has to be sampled in a limited resolution and ρ has to be quantified; (iii) $H(\rho, \theta)$ is also represented on a discrete grid where only integer coordinates have values; and (iv) the original image can suffer from various noises. These issues could cause problems of aliasing, peak spreading or peak extension [51–55].

However, it is difficult to directly use the HT method for images with high levels of noise, such as ultrasound images with significant speckle noises. Revoting Hough transform (RVHT) [56], an improved HT method, can identify the major muscle fascicles' orientations in musculoskeletal ultrasound images, which may contain significant speckle noises. RVHT uses a revoting strategy to deal with the aliasing problem in the line detection for ultrasound images and adopts a thresholding method using the data in HT space to control the detection of significant lines in ultrasound images.

The global maximum in the accumulator array, voted by all the edge/feature points, was first detected. Then all the feature points close to this line were removed from the edge map. With the updated edge map, the new accumulator array was computed and used to detect the global maximum by voting again. More lines could be identified by repeating this revoting procedure, as shown in Fig. 2.10. When the image was very noisy, the removal width could be extended from the basic value, 1 pixel, to several pixels (6 to 12 pixels were selected for the musculoskeletal ultrasound

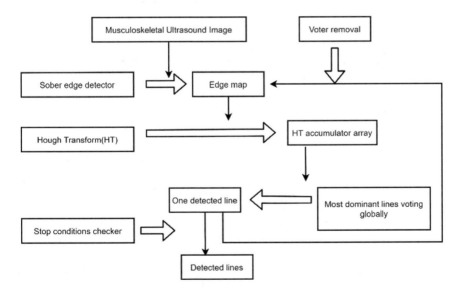

Fig. 2.10 Diagram of the procedures of the proposed revoting HT

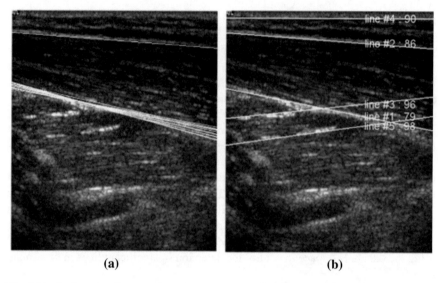

(a)   **(b)**

Fig. 2.11 Angle estimation results of a typical musculoskeletal ultrasound image. **a** the results using SHT with thresholds of 80% respectively, **b** lines detected using RVHT. Reprinted from Y. Zhou and Y.-P. Zheng, "Estimation of muscle fiber orientation in ultrasound images using revoting Hough transform (RVHT)," Ultrasound in medicine & biology, vol. 34, no. 9, pp. 1474–1481, 2008, with permission from Elsevier

images, with resolutions of 10 pixels/mm to 12.5 pixels/mm in this study) to remove more neighboring feature points along the location of the detected line. The center of the image was used as the origin of the coordinate system in HT, as suggested by Immerkær [50].

Compared with SHT, RVHT is inherently anti-aliasing as the feature points that voted for a line are removed when a line is detected (Fig. 2.11). As stated earlier, once the favored candidate line has won the global voting, voices of these feature points are ignored in the subsequent voting, and so are those in the nearby neighborhoods (depending on the removal width). If the image is very noisy, such as in many ultrasound images, manual detection becomes more subjective and difficult. In addition to the inherent subjective nature, RVHT is almost automatic and could potentially save lots of manual work in comparison with the traditional way of orientation estimation in musculoskeletal ultrasound images. Therefore, the proposed RVHT has potential for future image-guided ultrasound musculoskeletal analysis, especially for the automatic estimation of muscle thickness. The new method is particularly useful when continuous or real-time measurements are required, where manual detection is almost not possible because of the large number of images and the demand for a high processing speed [23, 13, 57–59].

The automatic MT estimation procedure, based on the RVHT method, could include four steps: (1) image enhancement, (2) thresholding operation to generate black-and-white images, (3) locating of superficial and deep aponeuroses by RVHT, and (4) computation of the distance between aponeuroses. RVHT first computes an accumulator matrix of Hough transform based on a black-and-white image called an "edge map." This edge map represents the meaningful image contents. The RVHT method then locates the global maximum in the accumulator matrix of Hough transform that corresponds to the most dominant line-shaped feature points globally, using the standard Hough transform. Then the pixels close to the detected line are removed from the edge map and the Hough transform accumulator matrix is calculated again. The same procedure can be executed to search for another line [56]. For ultrasound images of skeletal muscles, usually the very first two lines detected are the superficial and deep aponeuroses according to a priori knowledge. Finally, the mean distance between each two lines detected by RVHT is computed as the MT. A diagram of the procedures is shown in Fig. 2.12.

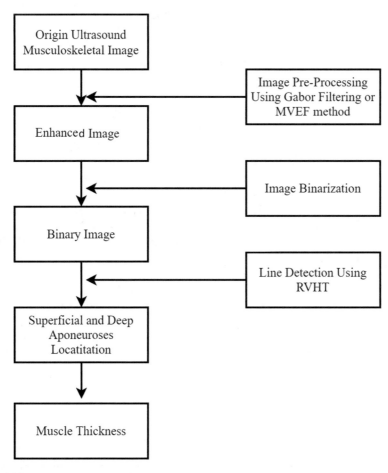

Fig. 2.12 The automatic MT estimation procedure

2.4.2 Pre-processing Techniques

The traditional manual operation to read muscle thickness is subjective and time-consuming; therefore, several studies have focused on the automatic estimation of muscle fascicle orientation and muscle thickness, to which the speckle noises in ultrasound images could be the major obstacle. There have been two popular methods proposed to enhance the hyperechoic regions over the speckles in ultrasonography, namely Gabor Filtering and Multiscale Vessel Enhancement Filtering (MVEF).

Taking into account that the fascicles and fibroadipose in a sonogram are tubular and include coherent orientation tendencies, the Gabor Filter bank method can be used to implement image enhancement. The method includes three steps: orientation filed estimation, frequency map computation and Gabor filtering confined by orientation reliability. More details can be found in the related studies [60, 61].

The Multiscale Vessel Enhancement Filtering method is based on the second-order local structure, with excellent noise and background suppression performance. The method also includes three steps: Hessian matrix estimation (including the choice of Gaussian kernels), computation of eigenvector for each scale, and processing for the maximum vesselness response. More details can be found in previous studies [62, 63].

The originally acquired images were first cropped to keep the image content only, and the cropped images were then enhanced by Gabor Filtering and MVEF methods respectively and binarized, as shown in Fig. 2.13. It should be noted that, in Gabor Filtering, empirically the default value of the reliability of the local orientation of image was set to 50%, which means that the corresponding location is ignored unless the reliability of the orientation is higher than 50%. Superficial and deep aponeuroses were then located by RVHT and MT was estimated automatically, as shown in Fig. 2.13b and c. After either Gabor Filtering or the MVEF method was used, superficial and deep aponeuroses were detected by RVHT as the very first two lines, as expected.

Without an enhancement procedure before RVHT, the performance of aponeuroses detection is quite poor. It is shown in a previous study [64] that Gabor Filtering and MVEF can both enable RVHT to generate comparable results of muscle thickness to those by manual drawing (mean $\pm$ SD, 1.45 $\pm$ 0.48 and 1.38 $\pm$ 0.56 mm respectively). However, the MVEF method requires much less computation than Gabor Filtering, suggesting that MVEF is more suitable for real-time applications where MT transitions are often of interest.

2.4.3 An Efficient Framework for MT Estimation

A new framework has been proposed for MT estimation in sonograms by skipping the voting step as much as possible [63]. Instead, the shape properties of the regions of interest are used to output the orientation in most cases. Compared with the single

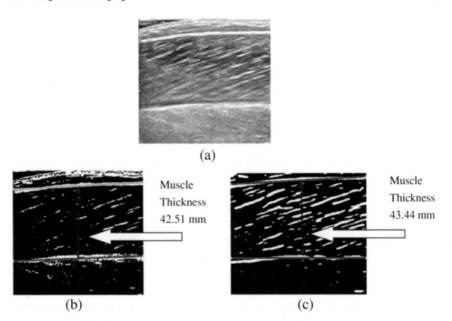

Fig. 2.13 Experimental results on a representative ultrasound image. **a** cropped image, **b** line detection using RVHT after Gabor Filtering, **c** line detection using RVHT after the MVEF method. The further MT estimation results are labeled in (**b**) and (**c**). Reprinted with permission from P. Han et al., "Automatic thickness estimation for skeletal muscle in ultrasonography: evaluation of two enhancement methods," Biomedical Engineering Online, vol. 12, no. 1, pp. 1–13, 2013. https://doi. org/10.1186/1475-925X-12-6

RVHT method, this framework is expected to achieve higher efficiency in regard to MT estimation without compromising the accuracy.

The efficient framework for estimation of MT includes three steps:

(1)   Image enhancement using the MVEF algorithm

Ultrasound images are usually affected by speckle noises, which hinder the analysis of musculoskeletal geometry. Taking into account the fact that the fibers in ultrasound images are tubular and include coherent orientation tendencies, MVEF enhances ultrasound images before aponeurosis detection. The MVEF method is based on the second-order local structure, with excellent noise and background suppression performance [63]. The method includes three steps: Hessian matrix estimation (including the choice of Gaussian kernels), computation of eigenvector for each scale, and processing for the maximum vesselness response. More details can be found in the previous study [63].

(2)   Extraction of object regions

Objects to be extracted from sonograms are regions which may represent long and thin muscle fibers. Since fibers, both aponeuroses and other fascia structures, have higher intensity than the background in ultrasound images, a straightforward approach to find the potential regions of fibers is to apply a threshold on the enhanced image, so that pixels whose value is greater than the threshold are regarded as pixels located in the potential object regions. In this realization, Otsu's method is employed to determine the optimal threshold [65]. This results in a binary map, where the white components represent the candidate regions for aponeurosis.

(3)   Estimation of MT

Shapes of the object regions of interest are found to vary and can be divided into three different patterns. RA: regions which are long and thin, where each represents one major muscle fiber. RB: regions which are long but have branches because of the adherence of two or several muscle fibers. RC: regions which are short and possibly form a 'broken' line caused by partial imaging of one single muscle fiber.

Locations of RA and RC are calculated by the ellipse that has the same normalized second central moments as the region. Locations of RB are calculated using Hough transform (HT). MT is calculated as the distance between aponeuroses. Specifically, three shape measures for the region, aspect ratio Ar, width ω, and length L, can be used for classification of RA, RB, and RC. In this study, L and ω are calculated respectively as the length of the major and minor axis of the ellipse that has the same normalized second central moments as the region, and the aspect ratio is defined as L/ω.

The procedures of the proposed framework for line detection on the binary map are shown in Fig. 2.14. A typical original ultrasound image of the biceps and the corresponding results obtained using the proposed method, are shown in Fig. 2.15. The lines and their angle values are marked with the order in which they were detected.

2.4.4   Registration-Based Method

All of the aforementioned methods are limited to the estimation of muscle thickness at one or several specific locations, assuming that the muscle thickness is constant longitudinally along the aponeuroses. But strictly, superficial and deep aponeuroses are not straight lines, nor always parallel to each other, which leads to changes of muscle thickness at different longitudinal locations of aponeuroses. Therefore, a new strategy is proposed to estimate muscle thickness for each point along the longitudinal axis of the ultrasound image based on extracting contours of aponeuroses, which is very useful when muscle thickness at more than one specific location needs to be detected.

The flowchart of the proposed strategy for muscle thickness measurement via ultrasound imaging is shown in Fig. 2.16. For each frame from the studied ultrasound image sequence, the first step is to acquire two seed points (one on the superficial aponeurosis, and the other on the deep aponeurosis). Then, contours of aponeuroses are extracted and muscle thickness is obtained by calculating the distance between contours of superficial and deep aponeuroses.

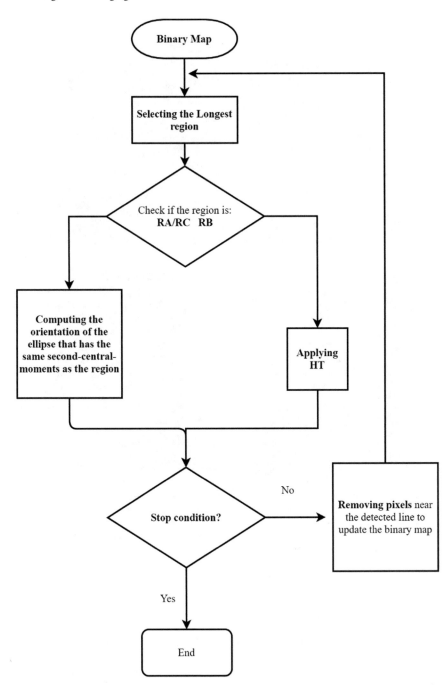

Fig. 2.14 Diagram of the efficient framework for aponeurosis detection

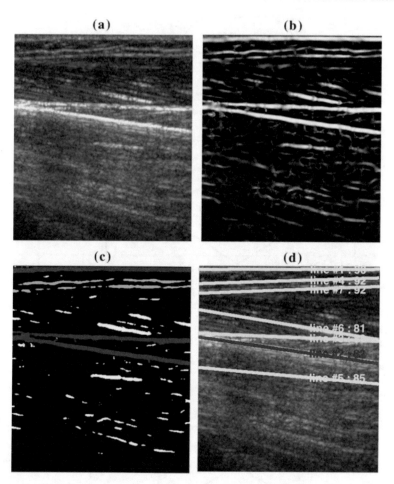

Fig. 2.15 A representative image of the biceps and line detection results. **a** The original image. **b** The image after MVEF. **c** The image after binarization, where red, blue, and green regions denote the typical RA, RB, and RC respectively. **d** line detection results, where the red line was detected using HT and the yellow lines were detected by the ellipse that has the same normalized second central moments as the region. Reprinted from Ling, S., Chen, B., Zhou, Y. et al. An efficient framework for estimation of muscle fiber orientation using ultrasonography. BioMed Eng OnLine 12, 98 (2013). https://doi.org/10.1186/1475-925X-12-98

(1)   Tracking of seed points

    Seed points in each frame are two points located at superficial and deep aponeuroses respectively, which are used to generate initial contours in the aponeuroses segmentation stage. Given changes of muscle shape and position in a sequence, locations of two seed points are adaptively adjusted in a frame-by-frame manner. However, considering the large number of images in a sequence, selecting seed points for every frame by hand is tedious and subjective. Here

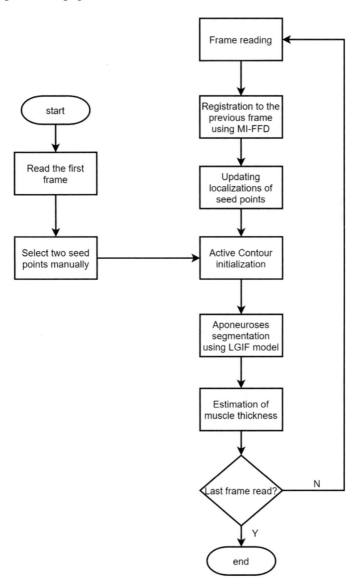

Fig. 2.16 Flowchart of the proposed strategy to measure MT from one ultrasound image sequence. (MI-FFD for mutual information-based free-form deformation; LGIF for local and global intensity fitting.) © 2021 IEEE. Reprinted, with permission, from S. Ling, Y. Zhou, Y. Chen, Y.-Q. Zhao, L. Wang, and Y.-P. Zheng, "Automatic tracking of aponeuroses and estimation of muscle thickness in ultrasonography: a feasibility study," IEEE Journal of Biomedical and Health Informatics, vol. 17, no. 6, pp. 1031–1038, 2013

an image registration method is adopted, named as mutual information-based free-form deformation (MI-FFD) [66] tracking, to automatically track seed points in the ultrasound image sequence. It produces local registration fields that are smooth, continuous and establish one-to-one correspondences. In this method, the transformation function which describes the deformation between two successive images is determined by minimizing a MI-based objective function.

In summary, for each trial, seed points of the first frame are determined manually as two arbitrary points inside the contours of aponeuroses. For two consecutive frames of each sequence, the previous frame is selected as a reference, and then, the MI-FFD method is applied to track the seed points in the subsequent frames automatically.

A representative example of the procedure for tracking of seed points is shown in Fig. 2.17. Figure 2.17a shows the current frame of one image sequence whose seed points have been determined, as represented by the two blue points in the figure. Seed points of the subsequent frame in the sequence are shown in Fig. 2.17b, where red points are true seed points determined by the MI-FFD method and blue points are at the same position as in Fig. 2.17a. It can be observed that the method can update positions of seed points back onto aponeuroses when the muscle contracts and the location of aponeuroses changes accordingly.

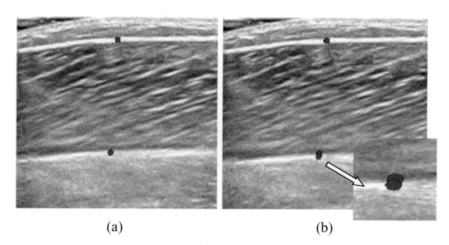

(a)   (b)

Fig. 2.17 Representative example of the procedure for tracking seed points. **a** One frame of a sequence with manually selected seed points is represented by two blue points. **b** Subsequent frame, where red points are true seed points determined by the MI-FFD method and blue points are at the same position as in (**a**). MI-FFD stands for mutual-information-based free-form deformation © 2021 IEEE. Reprinted, with permission, from S. Ling, Y. Zhou, Y. Chen, Y.-Q. Zhao, L. Wang, and Y.-P. Zheng, "Automatic tracking of aponeuroses and estimation of muscle thickness in ultrasonography: a feasibility study," IEEE Journal of Biomedical and Health Informatics, vol. 17, no. 6, pp. 1031–1038, 2013

(2) Aponeuroses segmentation

Active contour models have been widely used in image segmentation with promising results since its introduction by Kass et al. [67]. We adopted a region-based model, named as the local and global intensity fitting (LGIF) model [68], to extract contours of aponeuroses, with the advantages of accuracy and robustness. In this model, an energy function is defined with a local intensity fitting term, which induces a local force to attract the contour and stops it at object boundaries, and an auxiliary global intensity fitting term, which drives the motion of the contour far away from object boundaries. The combination of these two forces can handle intensity inhomogeneity and allows for flexible initialization of the contours. Denoting ϕ as the level set function, the energy functional of the model is defined as:

$$E(\phi) = (1 - \omega)E_{local}(\phi) + \omega E_{glocal}(\phi) + \nu L(\phi) + \mu P(\phi) \qquad (2.5)$$

where ω is a positive constant which determines the weight of the local force E_{local} and the global force E_{global}; $\nu > 0$ and $\mu > 0$ are constants as the weights of the length term $L(\phi)$ and the level set regularization term $P(\phi)$, respectively. Contours of each frame are initialized as two circles (radii $= 2$ pixels) centered at two seed points, which have been tracked by means of MI-FFD described previously. Then, entire contours of aponeuroses can be extracted by evolving the level set iteratively using the LGIF model. A representative example of the aponeuroses segmentation process using a LGIF evolution model is shown in Fig. 2.18.

(3) Estimation of muscle thickness

After segmentation of the aponeuroses, muscle thickness at each point along the aponeuroses is then estimated by calculating the distance between two corresponding points on superficial and deep aponeuroses, as shown in Fig. 2.19a. Muscle thickness of a representative frame estimated by the method is shown in Fig. 2.20.

Traditionally, muscle thickness is measured as the distance between two points, selected by an experienced operator, on the lower edge of a superficial aponeurosis and the upper edge of a deep aponeurosis respectively, as shown in Fig. 2.20. Therefore, in comparison to the manual method, the mean muscle thickness is estimated for each frame by calculating the mean distance between contours of superficial and deep aponeuroses, though this method outputs a curve of muscle thickness along the aponeuroses for each frame of an ultrasound image. A representative result of muscle thickness of the whole image sequence, measured by the proposed technique and the manual method, is displayed in Fig. 2.21.

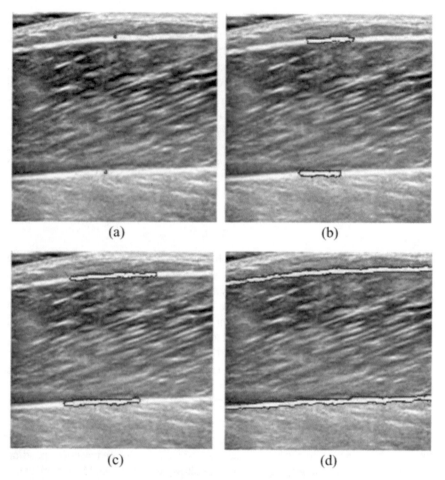

Fig. 2.18 Representative set of results of the LGIF model for aponeurosis segmentation. **a** Original image with initial contour. **b** Curve evolution result after 50 iterations. **c** Curve evolution result after 300 iterations. **d** Final contour after evolving stopped. © 2021 IEEE. Reprinted, with permission, from S. Ling, Y. Zhou, Y. Chen, Y.-Q. Zhao, L. Wang, and Y.-P. Zheng, "Automatic tracking of aponeuroses and estimation of muscle thickness in ultrasonography: a feasibility study," IEEE Journal of Biomedical and Health Informatics, vol. 17, no. 6, pp. 1031–1038, 2013

2.5 Summary

Muscle thickness measured using ultrasound images has many different applications, and it is also the first parameter used for sonomyography. Thickness SMG has been demonstrated to be closely correlated with muscle strength, and can be used for muscle fatigue assessment. It can be extracted from real-time ultrasound images in both transverse and longitudinal directions. Various techniques have been reported for the automatic detection of thickness SMG. In the future, muscle thickness SMG

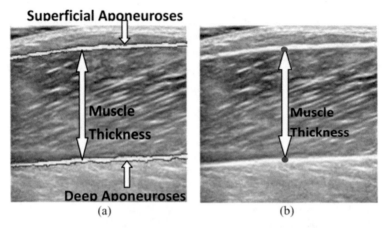

Fig. 2.19 Schematic diagrams of proposed strategy and manual method for muscle thickness measurement. **a** Strategy for muscle thickness measurement. Muscle thickness was estimated by calculating the average distance between contours of superficial and deep aponeuroses. **b** Manual method for muscle thickness measurement. Muscle thickness was approximated as the distance between two points on the lower edge of a superficial aponeurosis and upper edge of a deep aponeurosis, respectively. © 2021 IEEE. Reprinted, with permission, from S. Ling, Y. Zhou, Y. Chen, Y.-Q. Zhao, L. Wang, and Y.-P. Zheng, "Automatic tracking of aponeuroses and estimation of muscle thickness in ultrasonography: a feasibility study," IEEE Journal of Biomedical and Health Informatics, vol. 17, no. 6, pp. 1031–1038, 2013

Fig. 2.20 Muscle thickness estimated along the entire longitudinal axis of a representative frame

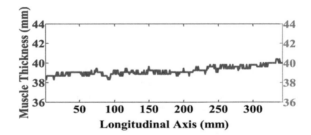

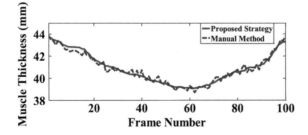

Fig. 2.21 Representative comparison result of muscle thickness measured by the proposed strategy and manual technique. © 2021 IEEE. Reprinted, with permission, from S. Ling, Y. Zhou, Y. Chen, Y.-Q. Zhao, L. Wang, and Y.-P. Zheng, "Automatic tracking of aponeuroses and estimation of muscle thickness in ultrasonography: a feasibility study," IEEE Journal of Biomedical and Health Informatics, vol. 17, no. 6, pp. 1031–1038, 2013

will have more potential applications particularly when the ultrasound hardware can be worn. Meanwhile, methods for the detection of SMG using deep learning and artificial intelligence can be adopted to achieve more reliable and fast performances.

References

1. Blumhagen, J., Noble, H.: Muscle thickness in hypertrophic pyloric stenosis: sonographic determination. Am. J. Roentgenol. **140**(2), 221–223 (1983)
2. Campbell, I.T., et al.: Muscle thickness, measured with ultrasound, may be an indicator of lean tissue wasting in multiple organ failure in the presence of edema. Am. J. Clin. Nutr. **62**(3), 533–539 (1995)
3. Teyhen, D. S., Rieger, J. L., Westrick, R.B., Miller, A.C., Molloy, J.M., Childs, J.D.: Changes in deep abdominal muscle thickness during common trunk-strengthening exercises using ultrasound imaging. J. Orthopaedic Sports Phys. Therapy **38**(10), 596–605
4. Wong, A., Gallagher, K.M., Callaghan, J.P.: Computerised system for measurement of muscle thickness based on ultrasonography. Comput. Methods Biomech. Biomed. Engin. **16**(3), 249–255 (2013)
5. Ichinose, Y., Kanehisa, H., Ito, M., Kawakami, Y., Fukunaga, T.: Morphological and functional differences in the elbow extensor muscle between highly trained male and female athletes. Eur. J. Appl. Physiol. **78**(2), 109–114 (1998)
6. Kubo, K., et al.: Muscle architectural characteristics in young and elderly men and women. Int. J. Sports Med. **24**(02), 125–130 (2003)
7. Freilich, R., Kirsner, R., Byrne, E.: Isometric strength and thickness relationships in human quadriceps muscle. Neuromuscul. Disord. **5**(5), 415–422 (1995)
8. Ikezoe, T., Asakawa, Y., Fukumoto, Y., Tsukagoshi, R., Ichihashi, N.: Associations of muscle stiffness and thickness with muscle strength and muscle power in elderly women. Geriatr. Gerontol. Int. **12**(1), 86–92 (2012)
9. Strasser, E.M., Draskovits, T., Praschak, M., Quittan, M., Graf, A.: Association between ultrasound measurements of muscle thickness, pennation angle, echogenicity and skeletal muscle strength in the elderly. Age **35**(6), 2377–2388 (2013)
10. Starkey, D.B., et al.: Effect of resistance training volume on strength and muscle thickness. Med. Sci. Sports Exerc. **28**, 10 (1996)
11. Miyatani, M., Kanehisa, H., Kuno, S., Nishijima, T., Fukunaga, T.: Validity of ultrasonograph muscle thickness measurements for estimating muscle volume of knee extensors in humans. Eur. J. Appl. Physiol. **86**(3), 203–208 (2002)
12. Miyatani, M., Kanehisa, H., Ito, M., Kawakami, Y., Fukunaga, T.: The accuracy of volume estimates using ultrasound muscle thickness measurements in different muscle groups. Eur. J. Appl. Physiol. **91**(2), 264–272 (2004)
13. Shi, J., Zheng, Y.-P., Chen, X., Huang, Q.-H.: Assessment of muscle fatigue using sonomyography: muscle thickness change detected from ultrasound images. Med. Eng. Phys. **29**(4), 472–479 (2007)
14. Ohata, K., Tsuboyama, T., Ichihashi, N., Minami, S.: Measurement of muscle thickness as quantitative muscle evaluation for adults with severe cerebral palsy. Phys. Ther. **86**(9), 1231–1239 (2006)
15. English, C.K., Thoirs, K.A., Fisher, L., McLennan, H., Bernhardt, J.: Ultrasound is a reliable measure of muscle thickness in acute stroke patients, for some, but not all anatomical sites: a study of the intra-rater reliability of muscle thickness measures in acute stroke patients. Ultrasound Med. Biol. **38**(3), 368–376 (2012)
16. Hadda, V., et al.: Intra-and inter-observer reliability of quadriceps muscle thickness measured with bedside ultrasonography by critical care physicians. Indian J. Critical Care Med.: Peer-Reviewed Official Publication Indian Soc. Critical Care Med. **21**(7), 448 (2017)

17. Bunce, S.M., Hough, A.D., Moore, A.P.: Measurement of abdominal muscle thickness using M-mode ultrasound imaging during functional activities. Man. Ther. **9**(1), 41–44 (2004)
18. Rabben, S.I., et al.: Semiautomatic contour detection in ultrasound M-mode images. Ultrasound Med. Biol. **26**(2), 287–296 (2000)
19. Abe, T., Kondo, M., Kawakami, Y., Fukunaga, T.: Prediction equations for body composition of Japanese adults by B-mode ultrasound. Am. J. Hum. Biol. **6**(2), 161–170 (1994)
20. Ishida, Y., Kanehisa, H., Carroll, J.F., Pollock, M.L., Graves, J.E., Leggett, S.H.: Body fat and muscle thickness distributions in untrained young females. Med. Sci. Sports Exerc. **27**(2), 270–274 (1995)
21. Miyatani, M., Kanehisa, H., Fukunaga, T.: Validity of bioelectrical impedance and ultrasonographic methods for estimating the muscle volume of the upper arm. Eur. J. Appl. Physiol. **82**(5), 391–396 (2000)
22. Segers, J., et al.: Assessment of quadriceps muscle mass with ultrasound in critically ill patients: intra-and inter-observer agreement and sensitivity. Intensive Care Med. **41**(3), 562–563 (2015)
23. Shi, J., Chang, Q., Zheng, Y.-P.: Feasibility of controlling prosthetic hand using sonomyography signal in real time: preliminary study (2010)
24. Xie, H.-B., Zheng, Y.-P., Guo, J.-Y., Chen, X., Shi, J.: Estimation of wrist angle from sonomyography using support vector machine and artificial neural network models. Med. Eng. Phys. **31**(3), 384–391 (2009)
25. Zheng, Y.-P., Chan, M., Shi, J., Chen, X., Huang, Q.-H.: Sonomyography: monitoring morphological changes of forearm muscles in actions with the feasibility for the control of powered prosthesis. Med. Eng. Phys. **28**(5), 405–415 (2006)
26. Cipriani, C., Zaccone, F., Micera, S., Carrozza, M.C.: On the shared control of an EMG-controlled prosthetic hand: analysis of user–prosthesis interaction. IEEE Trans. Rob. **24**(1), 170–184 (2008)
27. Chen, X., Zheng, Y.-P., Guo, J.-Y., Shi, J.: Sonomyography (SMG) control for powered prosthetic hand: a study with normal subjects. Ultrasound Med. Biol. **36**(7), 1076–1088 (2010)
28. Hokanson, D.E., Mozersky, D.J., Sumner, D., Strandness, D., Jr.: A phase-locked echo tracking system for recording arterial diameter changes in vivo. J. Appl. Physiol. **32**(5), 728–733 (1972)
29. Stadler, R.W., Taylor, J.A., Lees, R.S.: Comparison of B-mode, M-mode and echo-tracking methods for measurement of the arterial distension waveform. Ultrasound Med. Biol. **23**(6), 879–887 (1997)
30. Rana, M., Hamarneh, G., Wakeling, J.M.: Automated tracking of muscle fascicle orientation in B-mode ultrasound images. J. Biomech. **42**(13), 2068–2073 (2009)
31. Unser, M., Pelle, G., Brun, P., Eden, M.: Automated extraction of serial myocardial borders from M-mode echocardiograms. IEEE Trans. Med. Imaging **8**(1), 96–103 (1989)
32. Li, J., Zhou, Y., Lu, Y., Zhou, G., Wang, L., Zheng, Y.-P.: The sensitive and efficient detection of quadriceps muscle thickness changes in cross-sectional plane using ultrasonography: a feasibility investigation. IEEE J. Biomed. Health Inform. **18**(2), 628–635 (2013)
33. Doucet, A., De Freitas, N., Gordon, N.: An introduction to sequential Monte Carlo methods. In: Sequential Monte Carlo Methods in Practice. pp. 3–14. Springer (2001)
34. Cao, G., Jiang, J., Chen, J.: An improved object tracking algorithm based on image correlation. In: 2003 IEEE International Symposium on Industrial Electronics (Cat. No. 03TH8692), vol. 1, pp. 598–601. IEEE (2003)
35. Koo, T.K., Wong, C., Zheng, Y.: Reliability of sonomyography for pectoralis major thickness measurement. J. Manipulative Physiol. Ther. **33**(5), 386–394 (2010)
36. Loram, I.D., Maganaris, C.N., Lakie, M.: Use of ultrasound to make noninvasive in vivo measurement of continuous changes in human muscle contractile length. J. Appl. Physiol. **100**(4), 1311–1323 (2006)
37. Isard, M., Blake, A.: Condensation—conditional density propagation for visual tracking. Int. J. Comput. Vision **29**(1), 5–28 (1998)
38. Zhou, S.K., Chellappa, R., Moghaddam, B.: Visual tracking and recognition using appearance-adaptive models in particle filters. IEEE Trans. Image Process. **13**(11), 1491–1506 (2004)

39. Black, M.J., Jepson, A.D.: Eigentracking: robust matching and tracking of articulated objects using a view-based representation. Int. J. Comput. Vision **26**(1), 63–84 (1998)

40. Ho, J., Lee, K.-C., Yang, M.-H., Kriegman, D.: Visual tracking using learned linear subspaces. In: Proceedings of the 2004 IEEE Computer Society Conference on Computer Vision and Pattern Recognition, CVPR 2004, vol. 1, pp. I-I. IEEE (2004)

41. Ross, D.A., Lim, J., Lin, R.-S., Yang, M.-H.: Incremental learning for robust visual tracking. Int. J. Comput. Vision **77**(1–3), 125–141 (2008)

42. Wright, J., Yang, A.Y., Ganesh, A., Sastry, S.S., Ma, Y.: Robust face recognition via sparse representation. IEEE Trans. Pattern Anal. Mach. Intell. **31**(2), 210–227 (2008)

43. Mei, X., Ling, H.: Robust visual tracking using ℓ 1 minimization. In: 2009 IEEE 12th International Conference on Computer Vision. pp. 1436–1443. IEEE

44. Mei, X., Ling, H., Wu, Y., Blasch, E., Bai, L.: Minimum error bounded efficient ℓ 1 tracker with occlusion detection. In: CVPR 2011, pp. 1257–1264. IEEE (2011)

45. Jia, X., Lu, H., Yang, M.-H.: Visual tracking via adaptive structural local sparse appearance model. In: 2012 IEEE Conference on computer vision and pattern recognition, pp. 1822–1829. IEEE (2012)

46. Liu, B., Yang, L., Huang, J., Meer, P., Gong, L., Kulikowski, C.: Robust and fast collaborative tracking with two stage sparse optimization. In: European Conference on Computer Vision, pp. 624–637. Springer (2010)

47. Wu, Y., Lim, J., Yang, M.-H.: Online object tracking: a benchmark. In: Proceedings of the IEEE conference on computer vision and pattern recognition pp. 2411–2418. (2013)

48. Hough, P. V.: Method and means for recognizing complex patterns. In: Google Patents (1962)

49. Duda, R.O., Hart, P.E.: Use of the Hough transformation to detect lines and curves in pictures. Commun. ACM **15**(1), 11–15 (1972)

50. Immerkær, J.: Some remarks on the straight line Hough transform. Pattern Recogn. Lett. **19**(12), 1133–1135 (1998)

51. Kiryati, N., Bruckstein, A.M.: Antialiasing the Hough transform. CVGIP: Graph. Models Image Process. **53**(3), 213–222 (1991)

52. Van Veen, T., Groen, F.C.: Discretization errors in the Hough transform. Pattern Recogn. **14**(1–6), 137–145 (1981)

53. Kiryati, N., Lindenbaum, M., Bruckstein, A.M.: Digital or analog Hough transform? Pattern Recogn. Lett. **12**(5), 291–297 (1991)

54. Yuen, S.Y., Ma, C.H.: An investigation of the nature of parameterization for the Hough transform. Pattern Recogn. **30**(6), 1009–1040 (1997)

55. Lam, W.C., Lam, L.T., Yuen, K.S., Leung, D.N.: An analysis on quantizing the Hough space. Pattern Recogn. Lett. **15**(11), 1127–1135 (1994)

56. Zhou, Y., Zheng, Y.-P.: Estimation of muscle fiber orientation in ultrasound images using revoting hough transform (RVHT). Ultrasound Med. Biol. **34**(9), 1474–1481 (2008)

57. Guo, J.-Y., Zheng, Y.-P., Huang, Q.-H., Chen, X.: Dynamic monitoring of forearm muscles using one-dimensional sonomyography system. (2008)

58. Chleboun, G.S., Busic, A.B., Graham, K.K., Stuckey, H.A.:Fascicle length change of the human tibialis anterior and vastus lateralis during walking. J. Orthopaedic Sports Phys. Therapy **37**(7), 372–379 (2007)

59. Lichtwark, G., Bougoulias, K., Wilson, A.: Muscle fascicle and series elastic element length changes along the length of the human gastrocnemius during walking and running. J. Biomech. **40**(1), 157–164 (2007)

60. Zhou, Y., Zheng, Y.-P.: Longitudinal enhancement of the hyperechoic regions in ultrasonography of muscles using a gabor filter bank approach: a preparation for semi-automatic muscle fiber orientation estimation. Ultrasound Med. Biol. **37**(4), 665–673 (2011)

61. Hong, L., Wan, Y., Jain, A.: Fingerprint image enhancement: algorithm and performance evaluation. IEEE Trans. Pattern Anal. Mach. Intell. **20**(8), 777–789 (1998)

62. Frangi, A.F., Niessen, W.J., Vincken, K.L., Viergever, M.A.: Multiscale vessel enhancement filtering. In: International Conference on Medical Image Computing and Computer-Assisted Intervention, pp. 130–137. Springer (1998)

63. Han, P., et al.: Automatic thickness estimation for skeletal muscle in ultrasonography: evaluation of two enhancement methods. Biomed. Eng. Online **12**(1), 1–13 (2013)
64. Otsu, N.: A threshold selection method from gray-level histograms. IEEE Trans. Syst. Man Cybern. **9**(1), 62–66 (1979)
65. Ling, S., Zhou, Y., Chen, Y., Zhao, Y.-Q., Wang, L., Zheng, Y.-P.: Automatic tracking of aponeuroses and estimation of muscle thickness in ultrasonography: a feasibility study. IEEE J. Biomed. Health Inform. **17**(6), 1031–1038 (2013)
66. Huang, X., Paragios, N., Metaxas, D.N.: Shape registration in implicit spaces using information theory and free form deformations. IEEE Trans. Pattern Anal. Mach. Intell. **28**(8), 1303–1318 (2006)
67. Kass, M., Witkin, A., Terzopoulos, D.: Snakes: active contour models. Int. J. Comput. Vision **1**(4), 321–331 (1988)
68. Kwah, L.K., Pinto, R.Z., Diong, J., Herbert, R.D.: Reliability and validity of ultrasound measurements of muscle fascicle length and pennation in humans: a systematic review. J. Appl. Physiol. **114**(6), 761–769 (2013)

Chapter 3
Measurement of Skeletal Muscle Pennation Angle

Abstract Pennation angle (PA) is an important indicator for skeletal muscle activity, and automatic calculation of PA has high practical value in fields such as sports medicine and rehabilitation engineering. This chapter mainly introduces the automatic calculation methods used in PA measurement, especially focusing on those related to Hough transform and Radon transform, and their derivative versions designed to be more robust and accurate as well. Today, on different datasets, these methods have generally demonstrated calculation accuracy of PA measurement comparable to traditional manual methods, but are faster and more objective.

Skeletal muscles are voluntary active tissues in the human body which control actions and stabilize the skeleton [1]. It has been found that the force exerted by muscle fibers is modified by their geometric arrangement, structures of joints, and the angle and location of tendons in respect to bones, before such force appears at joints as moments (joint torque) in skeletal muscles. Thus the sole observation of joint actions gives little information on actions within muscle. For instance, the relationship between joint position and torque does not disclose the length–force characteristics of the muscle. This is especially true for pennate muscles in which short fibers are arranged at an angle to the line of action of the muscle, and the angle as well as muscle fiber length change during contraction [2]. Therefore, knowledge of the muscle architecture, expressed as the geometric arrangement of muscle fascicle, is important when studying muscle functions and the resultant joint actions [3]. Pennation angle (PA), defined as the acute angle between the fascicle orientation and deep aponeurosis orientation, is one of the most widely employed parameters in quantitative studies on skeletal muscles to reflect the state of muscle architecture [4, 5].

The pennation angle affects both the force production and the excursion (length change), and the fascicle length reflects the muscle contraction directly [6]. For identical muscular compositions and volumes, the longer the fascicles the higher the excursion and velocity of contraction. The larger the PA, the higher the contractile force potential [7] and the less the tension and the velocity of shortening potential of a given fiber within a muscle with respect to tendons [8].

Several studies have shown that isometric contraction alters the length and PA of muscle fascicles. As mentioned above, the magnitude of these changes in a single

© Springer Nature Singapore Pte Ltd. 2021
Y. Zhou and Y.-P. Zheng, *Sonomyography*, Series in BioEngineering,
https://doi.org/10.1007/978-981-16-7140-1_3

static contraction is determined by the force elicited in the muscle and the compliance of the in-series tendon. The higher the contractile force and the more compliant the tendon, the higher the fascicular shortening and PA increase with respect to rest. If, however, the same muscle is called on to contract repeatedly, its fascicular geometry during contraction may also be affected by the time-dependent properties of tendons. Numerous experiments have shown that tendons exhibit creep (i.e., they elongate over time) when loaded in an oscillating pattern, which suggests that repeated contractions might result in greater fascicular shortening and PA increase compared with a single contraction, thus altering the muscle's potential for force and joint moment production [7]. This study also demonstrates the importance of dynamic and accurate measurement of PA which can provide reliable evidence for continuous quantitative expression of skeletal muscle.

3.1 Manual Measurement

In the first instance, many of the studies on human muscle architecture have been performed through direct dissection of cadaver specimens. The typical process is described in previous studies [8, 9] and can be summarized as follows. The human hemipelvectomy sections are first fixed in formalin with the hip and knee joints in maximum extension and the ankle in maximum plantar flexion. As the muscles are dissected from the limbs, muscle lengths are measured as the distance between the most proximal and the most distal muscle fibers. The muscles are cleaned of fat and excessive connective tissue and weighed. The muscles are then prepared for architectural determinations as described by Sacks and Roy [10]. Briefly, individual muscles are put in a 10% formalin bath for two to three days, cleared in a sodium phosphate buffer, and then placed in 15% sulfuric acid to soften and digest the connective tissue. Subsequently, the muscles are cleared in phosphate buffer and stored in a 50% glycerol solution. The angle formed by the individual muscle fibers with the line of force exerted by the muscle is approximated with a protractor [11]. Bundles of 10–20 muscle fibers are dissected from several regions of each muscle, and the length of each bundle is measured to the nearest millimeter. Single fibers are dissected from selected bundles of each muscle and mounted on glass slides in glycerine jelly. Average sarcomere lengths are determined for each fiber under a light microscope (400X) by using a calibrated eyepiece micrometer. However, muscles in the embalmed cadavers have been reported to undergo changes of their morphological characteristics because of factors such as shrinkage [8]. Besides, this method does not allow one to study the effect of muscle contraction and changes in joint position on muscle architecture. Accurate quantification of the architectural changes of PA during muscle dynamic contraction is significant to understand the muscle function.

With improvements in the performance of computers, the imaging of muscles is widely applied in the analysis of their function. Muscle imaging is essential as it comprehends the biological characteristics and estimates the pathological conditions of muscle through observations of its architectural changes [12, 13]. Ultrasound imaging is a useful technique which provides non-invasive, non-radiative, low-cost and real-time screening of both normal and pathological muscle tissues [14]. Because of the ability to measure muscles in any position, and provide real-time dynamic images of muscle fascicle changes during muscle contraction, two-dimensional ultrasound images have been developed to study the morphological changes in muscle movements, including PA, and serve as a useful tool for clinical diagnosis and rehabilitation assessment.

Traditionally, the lines and angles in musculoskeletal ultrasound images were detected manually, or interactively using software such as NIH Image (National Institutes of Health, Bethesda, MD, USA; http://rsb.info.nih.gov/nih-image). Using the software, the orientation of a manually drawn line on the studied image could be read. These methods are very time-consuming and the manual detecting process is subjective, greatly affecting the wider applications of this parameter, particularly for the study of dynamic muscle contraction. However, the advancement of technology does not stop there. As the performance of embedded devices and the field of computer vision are developing rapidly, image processing methods become an effective means for dynamic measurement of PA. In the following sections, methods that are commonly used for the measurement of PA are introduced.

3.2 Hough Transform (RVHT)

The detection of a straight line in an image is a fundamental problem in computer vision. The Hough transform (HT) [15] of slope–intercept parameterization and its improved version, HT of angle–radius parameterization [16, 17], is an established solution to the problem. Because of the development and extension of HT technology, many studies have focused on the visual recognition tasks [18–20]. However, it is difficult to directly use these methods for images with high noise levels, such as ultrasound images with significant speckle noises.

In 2008, Zhou et al. [21] presented an improved HT method to identify the major muscle fascicle orientations in musculoskeletal ultrasound images, which may contain significant speckle noises and proposed a revoting strategy to deal with the aliasing problem in the line detection for ultrasound images. A thresholding method using the data in HT space was proposed to control the detection of significant lines in ultrasound images. They tested the performance of this new algorithm using computer-generated images with different levels of speckle noises and clinical musculoskeletal ultrasound images. The details of this method have been described in Sect. 2.4.1, where the measurement of muscle thickness is elaborated.

Standard Hough transform uses the normal parameterization of a straight line in an image. For ultrasound images of muscle, we aim to detect the fascicles, i.e. some specific line patterns. To test the feasibility of using HT methods for line orientation detection in musculoskeletal ultrasound images that are usually degraded by speckle

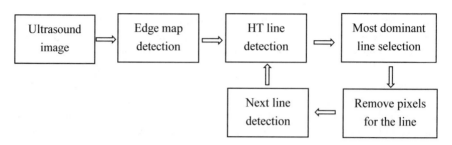

Fig. 3.1 Block diagram of the main procedures of the proposed revoting HT

noises, a modified HT named as revoting Hough transform (RVHT) was adopted. The main idea is to detect lines one by one, based on their dominance in images. After one line is detected, the pixels related to this line or those along the line will be removed, and then another line is detected. In this way, the dominant lines will affect the subsequent line detection, as lines in muscle ultrasound images are not sharp. The concept of this revoting procedure is shown in Fig. 3.1.

For the comparison of SHT and RVHT using a computer-generated image, a testing image (256 × 256), as shown in Fig. 3.2a, was generated using MATLAB 6.5 (The MathWorks, Inc., Natick, MA, USA). The distances from the vertexes of the larger equilateral triangle (with a side length of 200 pixels) to the left, right and upper borders of the image are all 28 pixels, whereas the distances from the vertexes of the smaller equilateral triangle (with a side length of 100 pixels) to the borders of the image are all 78 pixels. The base sides of both triangles are in the horizontal direction. The gray level between the two equilateral triangles was uniformly set to 64 and the background gray level was 0. Next, a new image (Fig. 3.2b) was created by adding a Gaussian noise with a mean of 64 (gray level) and a standard deviation of 16 to the original image. The edge map of Fig. 3.2b, acquired using the Sobel edge detector, is shown in Fig. 3.2c. Using all edge pixels in Fig. 3.2c, the accumulator array was then generated as shown in Fig. 3.2d, where the x-axis stands for θ (range 0–360) and y-axis for ρ (range 0 to the image diagonal length, 362 in this image); the labels for accumulator arrays had the same setting and therefore were omitted. Because in this implementation, the "angle" is defined as the angle between the detected line and the vector pointing from the image upper-left corner to the bottom-left corner, for one single straight line, two angles were detected, i.e., θ and its inverted version along the same straight line, $\theta + 180$. Subsequently, the 12 brightness peaks in Fig. 3.2d corresponded to the six vertexes of the two triangles, respectively. The specific correspondences will be further illustrated later after being removed one by one from the most-voted one. For the purpose of a better display, the accumulator array brightness was rescaled in each iteration, i.e., the current most voted points were also displayed as 22 Gy levels; therefore, Fig. 3.2i seems to have more sinusoidal curves than Fig. 3.2d, but it is not the case. It is only because of the rescaling of the displayed gray range after pixel removal was changed. The resolution of (ρ, θ) image is 1 degree in the x-axis and 1 pixel in the y-axis, which could satisfy

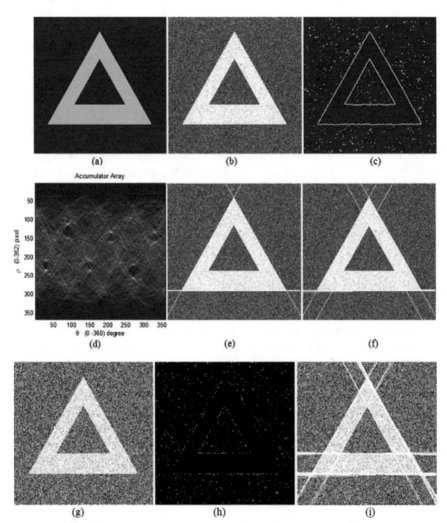

Fig. 3.2 Line detection results of a computer-generated image, using SHT [21]. **a** The original image (256 × 256); **b** the image after adding Gaussian noise, with a mean of 64 (gray level) and a standard deviation of 16; **c** the edge map acquired using the Sobel edge detector; **d** the accumulator array; **e** the line detection results when the SHT threshold is set to 70% of the global maximum; **f** the results obtained with a threshold of 50% of the global maximum; **g** the noisier image obtained by adding to the original image Gaussian white noise, with a mean of 64 and a standard deviation of 32; **h** the edge map of (g); and **i** the line detection results using SHT with the threshold set to 40% of the global maximum. In the accumulator array image, gray levels 0 and 255 represent its minimum and maximum values. Reprinted from Y. Zhou and Y.-P. Zheng, "Estimation of muscle fiber orientation in ultrasound images using revoting Hough transform (RVHT)," Ultrasound in Medicine & Biology, vol. 34, no. 9, pp. 1474–1481, 2008, with permission from Elsevier

Yuen's quantization schemes [22]. When the SHT threshold was set to 70% of the global maximum in the image, only the larger triangle was detected, as indicated in Fig. 3.2e. And when the threshold was decreased to 50% of the global maximum to make the Hough voting procedure more sensitive, two sides of the inner triangle could be detected (Fig. 3.2f). However, in this case, some extra lines were also detected, whereas the bottom side of the inner triangle was undetected, although it is obvious in the image. The main reason was that some lines generated large values, i.e., bright region, in the accumulated array image. The maximum value was larger than double the value generated by the bottom side of the inner triangle. To test a noisier case, Gaussian white noise with a mean of 64 and a standard deviation of 32 was added to Fig. 3.2a, producing a new image of Fig. 3.2g. Following the procedures mentioned earlier, the corresponding new edge map was obtained (Fig. 3.2h). Under this noise condition, the right side of the smaller triangle could not be detected with a threshold value of 50% of the global maximum in (ρ, θ) image. In addition, some extra lines could be observed near the base side of the larger triangle. When the threshold value was further decreased to 40%, more extra lines were detected (Fig. 3.2i), although all the required lines were detected.

As a comparison, the results of applying RVHT to the image of Fig. 3.2g are presented in Fig. 3.3. Figure 3.3a shows the obtained accumulator array, where the marked locations correspond to the two directions of the same straight line. The detected angle, defined as the angle between the detected line and the vector pointing from the image upper-left corner to the bottom-left corner as in the software of NIH Image [23], is 150/330°, which is correct according to the value used for the image generation. Figure 3.3b shows the "original image" (Fig. 3.2g) overlapped by the detected line and Fig. 3.3c the updated edge map where the edge/feature points corresponding to the detected line had been removed, with a removal width of 2 pixels. Similarly, using RVHT, 30/210° and 90/270° directions of the larger triangle, and 90/270°, 150/330° and 30/210° of the smaller triangle could be detected one by one, and the final results are shown in Fig. 3.3e.

For the application of RVHT on musculoskeletal ultrasound images, 45 ultrasound images of biceps and forearm muscles from three healthy adult male volunteers were acquired by an ultrasound image system (ATL HDI 5000, Philips Inc., Bothell, WA, USA) and cropped to remove the imaging tags and retain only the image content. In total, 168 lines were detected using RVHT, among which 165 were regarded as being valid according to visual verification. A typical image and the corresponding products using RVHT are shown in Fig. 3.4, where (a) is the original image, (b) its Sobel edge map, (c) the HT accumulator array of (b) and (f) the detected lines after RVHT. The line detection results using SHT with thresholds of 80% and 60% are shown in Fig. 3.4d, e, respectively. Considering the complicated speckle noise conditions in the ultrasound images, the peaks in the corresponding HT accumulator array image were no longer as "sharp" as those of the simulated image, as shown in Fig. 3.4c. In such case, the removal width for RVHT could be set to 12 pixels and the angles detected were marked in the figure in the order in which they were detected.

From the results obtained in the experiments, it was observed that the proposed RVHT has some advantages for line–angle detection in noisy images. Compared

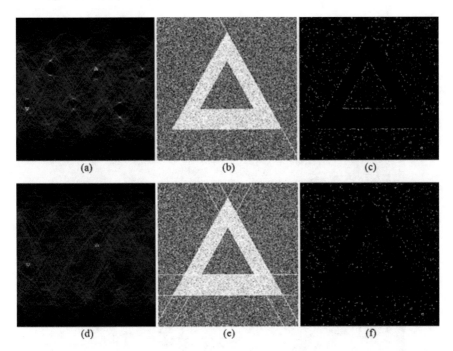

Fig. 3.3 Line detection results for the image of Fig. 3.2g using RVHT [21]. **a** The accumulator array, with the global maximum locations marked out, which correspond to the two directions of the same straight line; **b** the first detected line; **c** the updated edge map after the feature points corresponding to the first line were removed; **d** the HT accumulator array obtained after the feature points related to the first five lines were removed; **e** the last detected line; and **f** the updated edge map. In total, six lines were detected. In each accumulator array image, gray levels 0 and 255 represent its minimum and maximum values. Reprinted from Y. Zhou and Y.-P. Zheng, "Estimation of muscle fiber orientation in ultrasound images using revoting Hough transform (RVHT)," Ultrasound in Medicine & Biology, vol. 34, no. 9, pp. 1474–1481, 2008, with permission from Elsevier

with SHT, RVHT is inherently anti-aliasing as the feature points that had voted for a line were removed when the line was detected. As stated earlier, once the favored candidate line won the global voting, voices of these feature points were ignored in the subsequent voting, and so were those in the nearby neighborhoods (depending on the removal width). If the image was very noisy, such as in many ultrasound images, manual detection became more subjective and difficult. As demonstrated in this study, the proposed RVHT method could detect the lines with a performance comparable to that of manual operation ($R^2 = 0.965$), which used the mean of the results obtained by the two operators. In this study, we noted that the correlation between the results obtained by the two operators for manual detection ($R^2 = 0.9571$) was slightly lower than that between the RVHT result and the mean of manual results. The results indicate that there are some deviations between the results obtained by the two operators, although the majority of the results are well matched. Further studies are required to better understand the subjectivities of manual detection, with

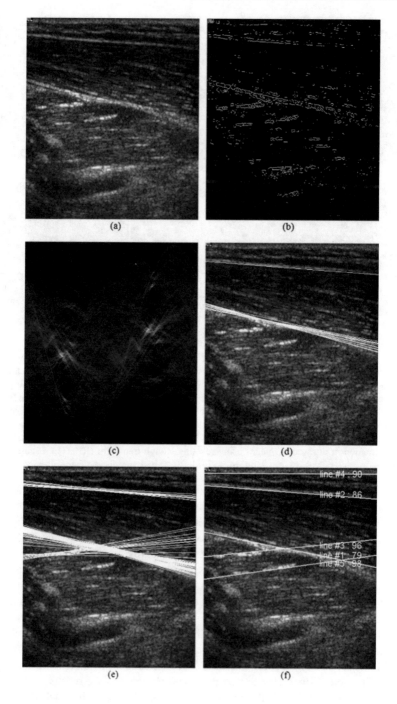

(a)   (b)

(c)   (d)

(e)   (f)

◀**Fig. 3.4** Angle estimation results of a typical musculoskeletal ultrasound image [21]. **a** The original image, **b** the Sobel edge map, **c** the accumulator array, **d** and **e** the results using SHT with thresholds of 80% and 60% respectively and **f** lines detected using RVHT; five lines in total were detected and their angles (between the detected line and the vector pointing from the upper-left to bottom-left corner of the image) were marked out. The removal width for RVHT was 12 pixels, and the ratio threshold of the feature points (counted by number of pixels in the edge map) of the last and first lines was set to 25%. Reprinted from Y. Zhou and Y.-P. Zheng, "Estimation of muscle fiber orientation in ultrasound images using revoting Hough transform (RVHT)," Ultrasound in Medicine & Biology, vol. 34, no. 9, pp. 1474–1481, 2008, with permission from Elsevier

more images having different noise conditions and more operators having different experiences. In addition to its inherent subjective nature, RVHT is almost automatic and can potentially save lots of manual work in comparison with the traditional way of orientation estimation in musculoskeletal ultrasound images. Therefore, the proposed RVHT has potential for future image-guided USI musculoskeletal analysis, especially for the automatic estimation of muscle thickness and fascicle length. The new method is particularly useful when continuous or real-time measurements are required, where manual detection is almost not possible because of the large amount of images and the demand for a high processing speed.

In spite of these advantages, the proposed RVHT could be improved in a number of aspects. Similar to the selection of the threshold value in SHT, RVHT faces a question of when to stop the revoting procedure. There are two possible solutions: (i) if the number of lines to be detected is given, the program can be stopped when that number is accomplished and (ii) if we know the feature differences between the expected lines and the unexpected ones, a threshold can be set according to the feature differences. In this study, we used the ratio of the number of feature points in the last and first detected lines to control the algorithm. This ratio was set to 25% in this study. We found that the larger this ratio, the better the feature points map of the image and the less time the algorithm consumes. The adaptive selection criteria of this ratio should be further investigated in future studies. This is particularly important for the real-time detection of pennation angles in the sonomyography technique, where the pattern of ultrasound images may change significantly during muscle contractions.

3.3 Radon Transform (RT)

As mentioned in Sect. 3.2, although the revoting strategy improves the aliasing problem in noisy ultrasound images, the method still greatly relies on the performance of the edge detector that could be compromised by speckle noise. However, high-intensity speckles may also be recognized and enhanced as small vessel structures. Radon transform (RT) has been widely used to detect linear features in synthetic-aperture radar (SAR) images [24–27] that have a comparable speckle noise level to ultrasound images. Compared to Hough transform (HT), RT does not require edge detection and its inherent integration feature is less susceptible to background noise.

Zhao et al. [28] presented a modified RT method that utilizes physiological orientations and locations as prior knowledge to identify muscle fascicles in musculoskeletal
ultrasound images. They also adopted the same revoting strategy as in RVHT to
extract lines sequentially. To track a fascicle continuously in a series of ultrasound
images, the position and orientation ranges where tracking is performed are updated
using the results from the previous frame. The proposed method is further validated
using both computer simulation and clinical data. The results show that the proposed
method has good performance in detecting and tracking fascicles in images with
blurry edges and low contrast. The procedures of the proposed methods are described
in detail as follows.

RT was first established in 1917 [29], nearly half a century before HT was introduced, which was found actually as just a special case of RT. The standard RT over
a 2-D Euclidean space is defined as:

$$f(\theta, \rho) = R\{F\} = \iint_D F(x, y)\delta(\rho - x \cos \theta - y \sin \theta)dxdy \qquad (3.1)$$

where D is the entire x–y image plane, $F(x, y)$ is the image intensity at position (x, y),
δ is the Dirac delta function, ρ is the distance from the origin (center of the image)
to the straight line, and θ is the angle between the x-axis and a line drawn from the
origin perpendicular to the straight line. According to the definition, RT accentuates
the linear features by integrating image intensity along all possible lines in an image,
so that the peaks in Radon space represent where the linear features are most likely
to be found.

But one of the drawbacks of RT is that intensity integration is performed over
the entire image. Therefore, fascicle segments significantly shorter than the whole
image dimensions are difficult to detect, especially when a high level of background
noise is present. In order to extract fascicles correctly, we introduced a localized
Radon transform (LRT) so that integration is performed within the proper range of
the locations and the directions where fascicles are supposed to be found. Thus,
Eq. 3.1 is modified as:

$$f(\theta, \rho) = R_{LOC}\{F\} = \int_{x_{min}}^{x_{max}} \int_{y_{min}}^{y_{max}} F(x, y)\delta(\rho - x \cos \theta - y \sin \theta)dxdy \qquad (3.2)$$

where only the points $x_{min} \leq x \leq x_{max}$ and $y_{min} \leq y \leq y_{max}$ in image space and θ_{min}
$\leq \theta \leq \theta_{max}$ in Radon space are calculated.

For fascicle detection, some prior knowledge can be used to localize RT. For
instance, fascicles have to be in between the superficial and deep aponeuroses that
usually have high echo intensity and are relatively easy to detect. The aponeuroses can
be extracted first over the entire image. Then, fascicles can be detected in the region
between the aponeuroses. The region can be further divided into smaller segments

until fascicles are detected correctly. It is also known that the pennation angle of the fascicles in a muscle is within a certain range (much smaller compared to the total range of 180°) even under pathological conditions. By using the possible fascicle orientations as prior knowledge, RT can be performed within a certain angle range instead of in every direction, so that only linear features that fall into the range can be detected.

After the image is transformed using LRT, the linear features are represented as peaks in Radon space. Sophisticated methods can be applied to suppress the background noise and separate lines that are closely located. For instance, regularization algorithms such as maximum entropy, l_1, l_ρ, and logarithm regularizations from an inverse problem perspective are used to suppress the nonlinear features and background noise, and a multiresolution line detection approach has been proposed to separate lines [30]. However, in Zhao's proposed method, after localization, the peaks are relatively easy to detect, and it is not necessary to extract all the fascicles simultaneously. Revoting strategy is a simple and robust way to extract line features one by one following the descending order of their integrated pixel intensities [20]. After the peak in Radon space is selected, the corresponding line in the image space is removed. Since in the ultrasound images fascicles do not intersect with each other, and usually there is a certain amount of space between fascicles, given a proper linewidth, all the feature pixels corresponding to the detected line can be removed while keeping other undetected lines intact. The same revoting and removing procedures are conducted until the specified number of detected lines has been reached, or the ratio between the peak amplitude and the average amplitude (peak-to-average) in Radon space is below a threshold, indicating that most significant linear features have been detected. In this study, the specified number of detected lines was used instead of peak-to-average threshold for better comparison between the LRT and RVHT methods.

Usually tracking of the same fascicle continuously over a series of images is more robust than detecting most prominent fascicles separately, because results from previous images can be used as prior knowledge to track the fascicle in the current image. In order to select the fascicle to be tracked, line detection as described earlier is performed to extract linear features from the first frame. As the fascicle is not supposed to translate or rotate much between two adjacent frames, RT is performed in the second frame within a relatively small region and limited directions around the fascicle detected in the first frame. Then, the peak is searched in the Radon space to detect the same fascicle in the second frame. The same procedure is applied to other frames by constantly updating the tracking region and direction range using the results from the previous frames.

In order to validate the performance of the LRT line detection method, simulation program Field II [31] was used to simulate ultrasound fascicle images. First, a phantom with the size of 40 mm × 40 mm × 10 mm (width × height × transverse width) was designed as shown in Fig. 3.5. A total of 100,000 scatterers were randomly (uniform) distributed within the phantom space. Since the thickness of aponeurosis ranges from 1 to 3 mm [32], two planes with 1-mm thickness were specified to simulate superficial and deep aponeuroses. The horizontal plane was 5 mm away from the

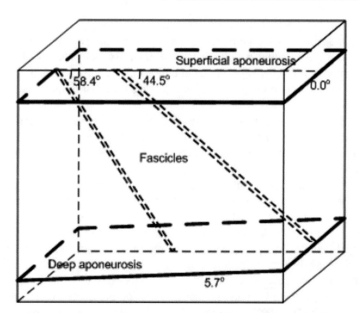

Fig. 3.5 Scatter phantom designed for Field II simulation [34]. A total of 100,000 scatterers were uniformly distributed in the space as illustrated in the figure. Two planes were placed in the phantom to simulate the superficial and deep aponeuroses. Two cylinders were placed between the two aponeuroses to simulate the fascicles. The angles of the aponeuroses and fascicles are displayed in the figure. ©2021 IEEE. Reprinted, with permission, from H. Zhao and L.-Q. Zhang, "Automatic tracking of muscle fascicles in ultrasound images using localized radon transform," IEEE Transactions on Biomedical Engineering, vol. 58, no. 7, pp. 2094–2101, 2011

surface of the phantom, while the oblique plane ranged from 2 to 6 mm away from the bottom of the phantom. The angles between the aponeuroses and the horizontal plane were 0° and 5.7°, respectively. In the center transverse plane, two cylinders with 1-mm radius [33] were placed between the two planes to simulate the fascicles. The angles between the fascicles and the horizontal plane were 44.5° and 58.4°, respectively. The echoes of the scatterers that fell into the simulated aponeuroses or fascicle regions were set ten times higher than the background echoes. The surface of the phantom was placed 30 mm away from the probe surface. A 128-element linear array transducer with the center frequency of 5 MHz and fractional bandwidth of 60% was designed to scan the phantom. The transmit focus of the transducer was 50 mm, and dynamic focusing at 30, 50, and 70 mm was applied to the echo signals. Hanning window apodization was applied to both transmit and receive. The system sampling rate was set at 100 MHz, and the frequency-dependent attenuation was 0.5 dB/(MHz cm). Echo signals of 50 scan lines were calculated using Field II, and the resulting image with 40 dB dynamic range is shown in Fig. 3.6.

Edge detector plays an important role in the RVHT method. Two popular detectors are the Sobel and Canny edge detectors. Compared to the Sobel edge detector, the Canny edge detector not only uses intensity gradient but also adopts a hysteresis

Fig. 3.6 Simulated phantom image with a dynamic range of 40 dB. The simulated aponeuroses and fascicles are clearly visible in the image. ©2021 IEEE. Reprinted, with permission, from H. Zhao and L.-Q. Zhang, "Automatic tracking of muscle fascicles in ultrasound images using localized radon transform," IEEE Transactions on Biomedical Engineering, vol. 58, no. 7, pp. 2094–2101, 2011

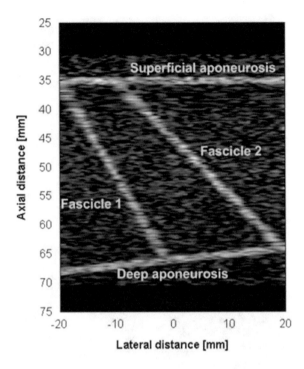

thresholding method to trace edges. Therefore, the Canny edge detector is less susceptible to speckle noise and able to trace faint sections of edges. However, in ultrasound images, lateral resolution is much lower than axial resolution, and the original images are usually excessively interpolated in the lateral direction to form B-mode images. Therefore, edges in the lateral direction become blurry and more difficult to track after interpolation.

Figure 3.6 was acquired by ten-time linear interpolation in the lateral direction, so that the number of pixels in the phantom image was the same in the lateral direction as in the axial direction. Some horizontal stripes due to interpolation are visible, as commonly observed in B-mode images. Figure 3.7a, b shows the edge maps of the phantom image using the Sobel and Canny edge detectors. It is difficult to see any proper edges using the Sobel edge detector, while by using the Canny edge detector the borders of the superficial and deep aponeuroses are visible. However, the edges of fascicles cannot be detected. Figure 3.7c, d shows the line detection results using RVHT based on (a) and (b) edge maps, respectively, overlapped with the original image. The entire edge maps are transformed into Hough space with 0.1° angle resolution, and the revoting procedure is applied to detect four lines that are supposed to be in the images. However, only superficial and deep aponeuroses are correctly extracted because their edges are more horizontal than the fascicles.

Applying the RT method to the same image in Fig. 3.7 with 0.1° angle resolution, four lines are sequentially detected, and results overlap the original image as shown

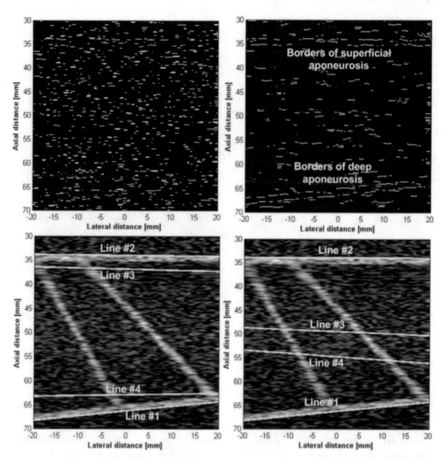

Fig. 3.7 Line detection based on RVHT using the simulated image in Fig. 3.6. **a** Edge map using the Sobel edge detector. **b** Edge map using the Canny edge detector. **c** Line detection result based on RVHT using (a). **d** Line detection result based on RVHT using (b). ©2021 IEEE. Reprinted, with permission, from H. Zhao and L.-Q. Zhang, "Automatic tracking of muscle fascicles in ultrasound images using localized radon transform," IEEE Transactions on Biomedical Engineering, vol. 58, no. 7, pp. 2094–2101, 2011

in Fig. 3.8. It can be seen that the aponeuroses and fascicles are correctly detected, because the extracted lines are determined by their integrated intensity no matter how blurry their edges are, and the measured angles are compared with theoretical ones in Table 3.1. The differences between the measured values and the phantom design values are within 0.5°, which is probably due to the error (scatterer distribution, focusing, beamforming, etc.) introduced by Field II.

To make the grayscale of line structures closer to the background, the same data were used to form a B-mode image with a 50 dB dynamic range that is shown in Fig. 3.9a, so that it becomes more difficult to extract the lines from the image. In this case, when RT is applied to the entire image to extract four most prominent

Fig. 3.8 Line detection result using LRT. Four lines are extracted sequentially from the image using a revoting technique. ©2021 IEEE. Reprinted, with permission, from H. Zhao and L.-Q. Zhang, "Automatic tracking of muscle fascicles in ultrasound images using localized radon transform," IEEE Transactions on Biomedical Engineering, vol. 58, no. 7, pp. 2094–2101, 2011

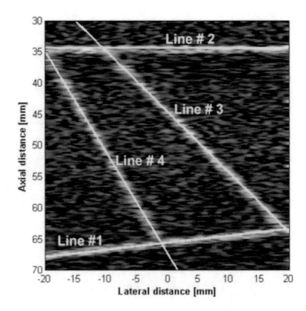

Table 3.1 Linear feature detection results using RT. ©2021 IEEE. Reprinted, with permission, from H. Zhao and L.-Q. Zhang, "Automatic tracking of muscle fascicles in ultrasound images using localized radon transform," IEEE Transactions on Biomedical Engineering, vol. 58, no. 7, pp. 2094–2101, 2011

	True (°)	Estimated results (°)
Deep apo	−5.7	−6.2
Superficial apo	0.0	0.1
Fascicle #1	44.5	44.0
Fascicle #2	58.4	58.2

line structures, only the two aponeuroses and fascicle #1 can be correctly detected, as shown in Fig. 3.9b. The fourth line has the most integrated intensity after three lines are removed. Suppose that the physiological angle range is between 20° and 90° which is much larger than the real range, LRT is performed, and the fascicles are correctly detected. Figure 3.9c shows the result after applying angle localization.

From the simulations, Zhao et al. suggested that the RVHT method relies on edge detectors that do not work well for linear features in ultrasound images [34]. Although HT is a discrete version of RT, RT is more intuitive in specifying the angle and region range where the linear features are to be detected. Judicious choice of localization strategies can be made based on the location and orientation of the fascicles in the image. Therefore, fascicle detection in a single image might be semiautomatic. However, localization techniques are extremely helpful in tracking fascicles over a series of images, which is shown as follows.

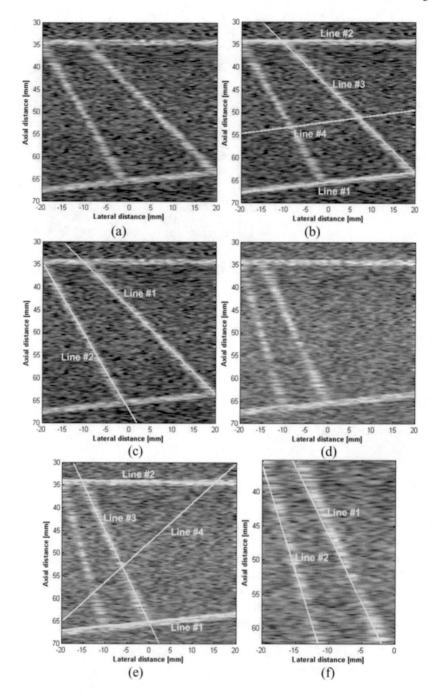

To further validate the performance of their proposed method, an ultrasound image of soleus (SOL) muscle fascicles was collected from a subject with a high body-mass index (BMI, 33.5 kg/m^2) using a commercial ultrasound imaging system, GE LOGIQ-9 with a 12 MHz probe M12 L (GE Healthcare, Waukesha, WI). The image is shown in Fig. 3.10a. The deep aponeurosis is invisible because of the higher backscattering from the fatty tissues. Only blurry outlines of two fascicles are partially observable in the image. RVHT with Canny edge detection was applied to the image and the resulting edge map and the line detection results are shown in Fig. 3.10b, c, respectively. Three lines were extracted. However, only the superficial aponeurosis and one fascicle was correctly detected. RT was applied to the same image and three lines were extracted as shown in Fig. 3.10d. The superficial aponeurosis and two fascicles are correctly detected. The results show that their proposed method performs better than RVHT in blurry images.

Another simple experiment was carried out to further test their proposed tracking method. A healthy subject was seated in a custom knee-ankle joint-driving device detailed in [35], with the right knee fixed at 0° (full extension) and the right ankle fixed in a footplate driven by a servomotor. The same ultrasound imaging system was used to obtain images of the medial gastrocnemius (MG) muscle. The probe was fixed to the leg using a probe holder where the MG fascicles could be clearly visualized. Then, the ankle was rotated back and forth from 30° plantar flexion to 20° dorsiflexion at a constant speed of 7.14°/s, while ultrasound video data were recorded at a frame rate of 17.1 frames/s. The ankle position was recorded on a PC that also sent out trigger signals to synchronize with the ultrasound data [36]. The ultrasound video data were converted into a series of images, and 239 consecutive frames corresponding to a cycle of moving the ankle from 30° plantar flexion to 20° dorsiflexion then back to 30° plantar flexion were selected to detect and track the fascicles using the proposed method.

The original image size was 532 × 432 as shown in Fig. 3.11a and a region of 350 × 300 was selected for each image where the fascicles and aponeuroses were dominant (see Fig. 3.11b). The direction range of the fascicles was roughly known to be between 0° and 20°, and the range of the aponeuroses was between 160° and 180°. First, the Canny edge detector was applied to the first frame, and the resulting edge map is shown in Fig. 3.11b. Edges of the aponeuroses and parts of some fascicles are visible. RVHT was performed using the edge map and seven lines were extracted as shown in Fig. 3.11c. The resolution of RVHT was 0.1° and the width of the lines to be eliminated in the revoting process was set at 10 pixels. As

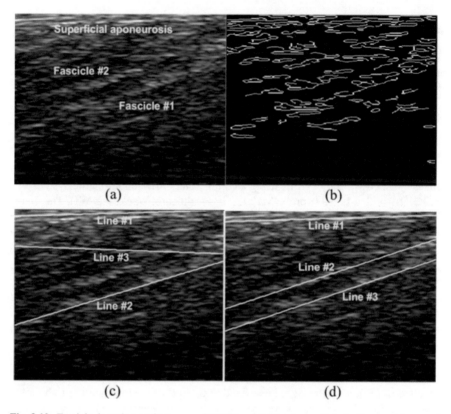

Fig. 3.10 Fascicle detection performance of RVHT and LRT using a clinical ultrasound image of SOL muscle fascicles from a subject with a high BMI. **a** Original SOL fascicle image. **b** Edge map generated using the Canny edge detector. **c** RVHT line detection result based on the edge map as shown in (b). **d** Line detection result using the RT method. ©2021 IEEE. Reprinted, with permission, from H. Zhao and L.-Q. Zhang, "Automatic tracking of muscle fascicles in ultrasound images using localized radon transform," IEEE Transactions on Biomedical Engineering, vol. 58, no. 7, pp. 2094–2101, 2011

shown in Fig. 3.11c, there are some discrepancies between the detected and original aponeuroses probably because the edges were not clearly defined, while lines #6 and #7 were falsely detected. In Fig. 3.11d, LRT was performed in the same region of the image to detect seven lines including two aponeuroses and five fascicles. The resolution of RT was still 0.1° and the width of the lines to be eliminated in the revoting process was set at 10 pixels. Then, the two aponeuroses and one fascicle were tracked separately throughout the 238 frames. The fascicle in each frame was tracked in the 10-pixel-wide region and 2° of direction range around the detection result in the previous frame. Since the aponeuroses were not supposed to move much during the process, they were tracked in the 5-pixel-wide region and 0.4° of direction range around the detection result in the previous frame. Representative

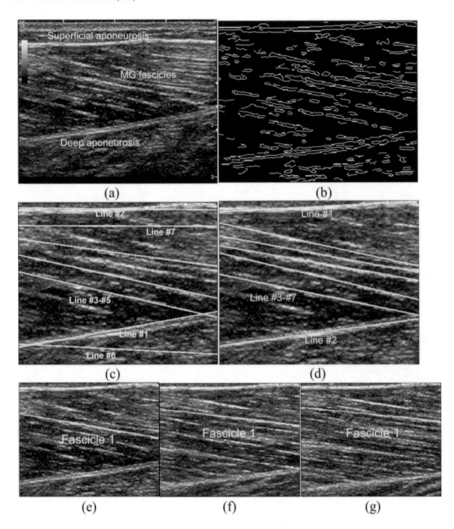

Fig. 3.11 Fascicle detection performance using LRT in clinical ultrasound images **a** Ultrasound image of an MG muscle from a healthy subject. **b** Part of the image is selected to detect edges using the Canny edge detector. **c** Linear feature detection result using RVHT with Canny edge detection. Of the seven detected lines only two of the fascicles are correctly extracted. **d** Linear feature detection result using LRT. Seven lines including superficial, deep aponeuroses, and five fascicles are correctly detected. One of the five fascicles is tracked using LRT throughout the 239 frames of ultrasound images as the ankle moves from 30° plantar to 20° dorsiflexion and back. Three representative tracking results of the same fascicle corresponding to 30° plantar flexion, 0°, and 20° dorsiflexion are shown in **e–g**, respectively. ©2021 IEEE. Reprinted, with permission, from H. Zhao and L.-Q. Zhang, "Automatic tracking of muscle fascicles in ultrasound images using localized radon transform," IEEE Transactions on Biomedical Engineering, vol. 58, no. 7, pp. 2094–2101, 2011

fascicle tracking results at frames 2, 72, and 120, corresponding to an ankle angle of 30° plantar flexion, 0° and 20° dorsiflexion, are shown in Fig. 3.11e–g, respectively.

Figure 3.12a shows the ankle position corresponding to each image frame. The fascicle orientation curve during ankle rotation, together with the superficial and deep aponeuroses is plotted in Fig. 3.12b. Then, pennation angles between fascicles and aponeuroses are calculated as shown in Fig. 3.12c. The same set of images was also analyzed by an experienced sonographer independently in every 5° ankle rotating angle. The curves were plotted and compared with results using the automatic method. The difference between the two methods was within 1°. According to the

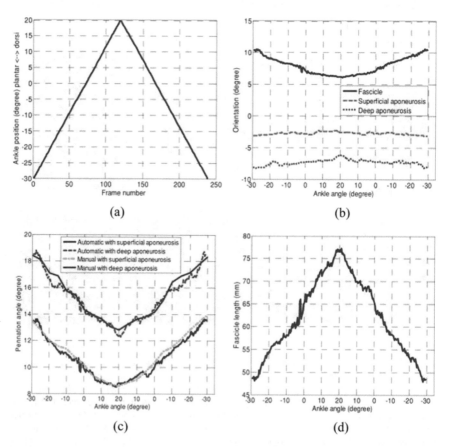

Fig. 3.12 Further tracking results of the experimental study. **a** Relation between the ankle position and the frame number. **b** Orientation–ankle angle curves of aponeuroses and the fascicle by tracking each of them separately. **c** Pennation angle–ankle angle curve of an MG muscle fascicle during ankle rotation. The results are compared with manual measurement every 5° during ankle rotation. **d** Estimated fascicle length–ankle angle curve of an MG muscle fascicle during ankle rotation. ©2021 IEEE. Reprinted, with permission, from H. Zhao and L.-Q. Zhang, "Automatic tracking of muscle fascicles in ultrasound images using localized radon transform," IEEE Transactions on Biomedical Engineering, vol. 58, no. 7, pp. 2094–2101, 2011

results, the pennation angle with superficial aponeurosis decreased from 13.6° to 8.5° and the pennation angle with deep aponeurosis decreased from 18.8° to 12.3°, as the ankle moved from 30° plantar to 20° dorsiflexion. Assuming that the MG muscle aponeuroses are straight and can be extended out of the image, the positions where the fascicle intersects with the aponeuroses can be determined. Therefore, the fascicle length is calculated as the distance between the intersections. The fascicle length curve in Fig. 3.12d indicates that the fascicle length increased from 48.6 to 77.6 mm as the ankle moved from 30° plantar to 20° dorsiflexion.

Although there are some fluctuations in both orientation and length results probably due to the irregular jitter between the muscle and the probe, the trend of the curves still coincides with the ankle positions. In addition, according to the results, both pennation angle and fascicle length change faster near plantar flexion extremity compared to dorsiflexion. The information on the dynamics of fascicles and aponeuroses during ankle joint rotation can be obtained that otherwise would be a great amount of work to manually measure multiple lines per frame for hundreds of frames.

In this section, we have described in detail the method of RVHT and Zhao's modified Radon transform, but it should be pointed out that the comparison of the two methods involves the scale of the dataset, the experimental design and parameter adjustments, and therefore it is up to the user to choose a suitable method under various circumstances.

3.4   Orientation of Fascicle Patterns

With substantial changes in pennation angle that occur with large joint movements and muscle contractions, the local features within the ultrasound image change in shape and may even disappear from the ultrasound plane and this can result in error in the measurement. Instead of measuring orientation of individual fascicle, some studies tried to find the dominant fascicle orientation. Besides the methods from Sect. 3.3, Rana et al. [36] developed two methods for automatically tracking fascicle orientation. Images were initially filtered using multiscale vessel enhancement (a technique used to enhance tube-like structures), and then fascicle orientations were quantified using either Radon transform or wavelet analysis to quantify the orientation of the fascicles within a muscle from ultrasound images, and to resolve the orientations to localized regions along the fascicles.

Specifically, B-mode ultrasound images contain information on muscle fascicle orientation as well as noise. A multistage process was used to determine fascicle orientation (Fig. 3.13), with an initial multiscale vessel enhancement filtering enhanced fascicle structure (which is vessel-like or tubular) and decreased noise level. Fascicle orientation was then determined by two alternative methods: **a** Radon transform was used to quantify the dominant orientation in the image and **b** an ultrasound-specific wavelet analysis quantified the local orientation around each

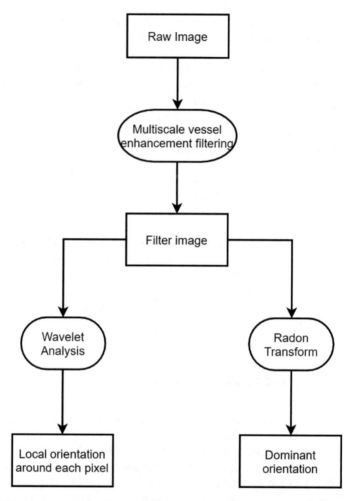

Fig. 3.13 The sequence of methods used to determine fascicle orientation. Reprinted from M. Rana, G. Hamarneh, and J. M. Wakeling, "Automated tracking of muscle fascicle orientation in B-mode ultrasound images," Journal of Biomechanics, vol. 42, no. 13, pp. 2068–2073, 2009, with permission from Elsevier

pixel. The two methods were validated against synthetic images with known orientation and also with real ultrasound images that were additionally digitized manually by 10 individuals.

Multiscale vessel enhancement filtering Muscle fascicles appear dark in the image and connective tissue between the fascicles appears as bright, vessel-like tubular structures that parallel the fascicles. Frangi et al. [37] used a vesselness filter to enhance muscle fascicles, which is also previous introduced in Sect. 3.4.2. An

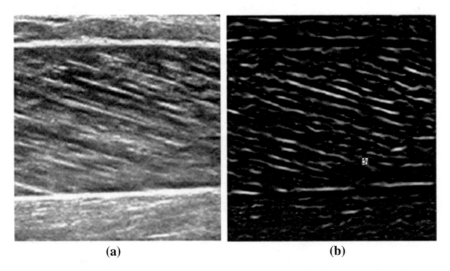

(a) **(b)**

Fig. 3.14 a Ultrasound image from the GM muscle, and **b** the image after multiscale vessel enhancement filtering

example image and its MVEF-processed version by Rana et al. [36] are shown in Fig. 3.14.

Anisotropic wavelet analysis Muscle fascicle orientation at each pixel was obtained by using anisotropic wavelet analysis. A wavelet kernel was constructed based on a modified Mortlet wavelet that was extended into 3D and given polarization with a major orientation a (Fig. 3.15). The kernel was 2k + 1 pixels in both the x and y directions. At any pixel the amplitude of the wavelet G(x, y) was given by:

$$G(x, y) = \exp\left(\frac{x^2 + y^2}{-dk}\right) \cos\left(\frac{2\pi(x \cos\alpha - y \sin\alpha)}{f}\right) + o, \qquad (3.3)$$

where d is the damping of the wavelet, f the spatial frequency, and o a linear offset. The spatial frequency should be set to a major frequency of the repeating fascicular structure; this will vary between the ultrasound equipment used but can be determined by using Radon transform on the image where the frequency of the fluctuations of the projection (Fig. 3.16c) is the same as the spatial frequency of lines in the image. In this study $f = 7$. The damping was set at $d = 2.5622$ to (a) provide decay of the wavelet by the edges of the kernel (with a half-width of $k = 20$) and (b) to satisfy the wavelet condition of zero integral (for $\alpha = 0$ and $o = 0$). However, due to pixilation artifacts, a non-zero value for o must be introduced for non-zero angles α in order to maintain a zero integral. The value for o never exceeded 0.0004% of the maximum value for the wavelet, and so this offset correction made a negligible difference to the results.

(a)   **(b)**

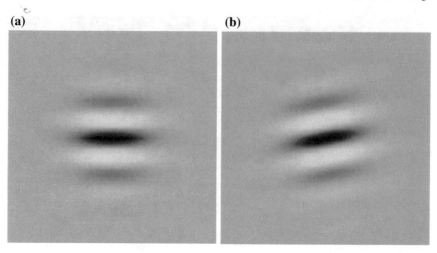

Fig. 3.15 Anisotropic wavelets for identifying fascicle direction within ultrasound images. Wavelets were calculated using Eq. (3.2), with $d = 2.5622$, $k = 20$, and $f = 7$ and formed a 41×41 pixel grid. Wavelets are shown for orientation $\alpha = 0°$ **a** and $\alpha = 10°$ **b**. Reprinted from M. Rana, G. Hamarneh, and J. M. Wakeling, "Automated tracking of muscle fascicle orientation in B-mode ultrasound images," Journal of Biomechanics, vol. 42, no. 13, pp. 2068–2073, 2009, with permission from Elsevier

The filtered image was convolved with a set of wavelet kernels at different orientations α. The wavelet with orientation α that resulted in the greatest convolution for a given region in the image identified that region as having a muscle fascicle orientation. This region as a 160×160 pixels grid was selected with its center in the middle of the x-axis of the image and half the distance between superficial and deep aponeuroses.

Radon transform As mentioned above, Radon transforms can be used to determine the predominant orientation in a repeated structure such as the muscle fascicles in an ultrasound image. Radon transform projects a grid of parallel lines, one pixel apart, across the image and calculates the integral of the image intensities along each line. The orientation y of the grid is varied, and when y approaches the dominant orientation of structures within the image then Radon transform has greatest variability across the image (Fig. 3.16), and this variability was quantified by its variance or kurtosis. Variance was used for synthetic images, as has been used earlier by Khouzani and Zadeh [38] and kurtosis was used for real ultrasound images because it was found that it worked better for real images than variance. Higher kurtosis indicates that more of the variance is due to infrequent extreme deviations, as opposed to frequent modestly-sized deviations. Since the real images have discontinuous line-like structures at unequal distances relative to each other (as opposed to uniformly distributed continuous lines in the synthetic images) the measure of kurtosis is more sensitive to fascicle orientation in the ultrasound images than the variance. The dominant fascicle

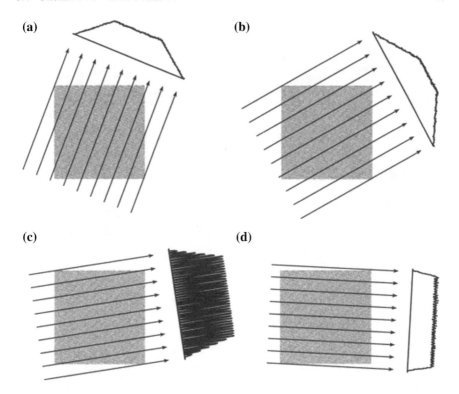

Fig. 3.16 An illustration of Radon transform at four different angles θ on a synthetic grid **a–d**. The arrows show the projections through the grid. Radon transform shows the greatest variance when θ approaches the orientation within the grid. Reprinted from M. Rana, G. Hamarneh, and J. M. Wakeling, "Automated tracking of muscle fascicle orientation in B-mode ultrasound images," Journal of Biomechanics, vol. 42, no. 13, pp. 2068–2073, 2009, with permission from Elsevier

orientation was taken as the angle y at which Radon transform of the filtered image had the greatest variance (synthetic grids) or kurtosis (ultrasound images).

The use of wavelet and Radon transforms for ultrasound images was validated using both synthetic and real images. Synthetic images were grids of parallel lines at a known orientation that had sinusoidal changes in intensity across (or perpendicular to) the lines, and these grids were combined with random noise (Fig. 3.17). A contrast to noise ratio (CNR) was defined as the ratio of the difference in intensity between bright and dark regions of the image to the sum of noise from those respective regions. For images with distributed pixel intensities, the difference in intensity between bright and dark regions was taken as the difference between the mean pixel intensities from the brightest and darkest 50% of the pixels. The noise from each region was quantified by the standard deviation of the image pixels in each region:

$$CNR = \frac{I_{Bright} - I_{Dark}}{\sigma_{Bright} + \sigma_{Dark}} \tag{3.4}$$

(a)   **(b)**

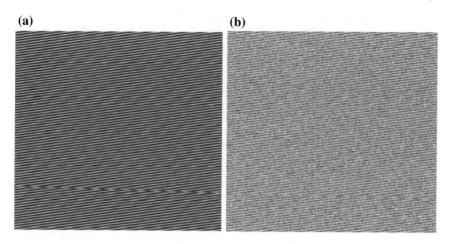

Fig. 3.17 Synthetic images used for validations at an orientation of 8.6° with no added noise **a** and with added noise at CNR = 0.79 **b**. Reprinted from M. Rana, G. Hamarneh, and J. M. Wakeling, "Automated tracking of muscle fascicle orientation in B-mode ultrasound images," Journal of Biomechanics, vol. 42, no. 13, pp. 2068–2073, 2009, with permission from Elsevier

The two methods were tested for synthetic grids at a fixed angle of 10° and different noise levels and then at fixed noise level (CNR = 1.0) but different angles. To validate the methods against real images, B-mode ultrasound images (Echoblaster, Telemed; LT) were recorded at 45 Hz from the distal part of the left vastus lateralis of a subject during cycling on a stationary ergometer. A linear-array probe (128 elements at 7 MHz) was secured to the skin with elasticated bandages and aligned to a plane in which the muscle fascicles were situated. Bitmap images were extracted for each frame from the ultrasound sequence. Sixty ultrasound images from the ultrasound sequence were manually digitized by 10 different researchers. Each person digitized the sequence twice, and only the second measurements were used to allow for a training effect in visualizing the fascicles. From each image, a fascicle was identified that spanned from superficial to deep aponeurosis, and two points were digitized on the fascicle close to the aponeurosis. The angle between these points on the fascicle and the x-axis was calculated. A region of interest was identified for each image within the aponeurosis that contained only the muscle fascicles, and this fascicle direction was quantified by the mean obtained from the wavelet analysis and the dominant orientation y from Radon transform (Fig. 3.18).

Rana et al. found that the wavelet analysis and Radon transform methods are able to accurately identify fascicle orientation [36] in synthetic images (Fig. 3.19) with a CNR of 0.8. The least squares linear regression for the calculated orientations against the actual orientations for these two methods both have slopes that are not significantly different from unity (the ideal line). The mean absolute error for Radon transform across the range of 0–90° was 0.058° and the mean error for wavelet transform across this range was 0.02°. These error values are small enough for practical applications.

Fig. 3.18 Ultrasound image from the vastus lateralis [36]. The aponeuroses are indicated by the dashed lines. The mean fascicle orientations are shown by the solid black and gray lines from Radon and wavelet transforms, respectively. Each line is shown against a white relief for clarity. Reprinted from M. Rana, G. Hamarneh, and J. M. Wakeling, "Automated tracking of muscle fascicle orientation in B-mode ultrasound images," Journal of Biomechanics, vol. 42, no. 13, pp. 2068–2073, 2009, with permission from Elsevier

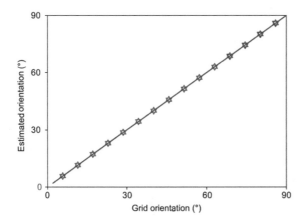

Fig. 3.19 Estimated orientations from the simulated grids calculated for a range of angles using CNR = 0.75. Angles calculated using wavelet transform are shown by triangles with an apex at the top, and angles calculated using Radon transform are shown by triangles with an apex at their bottom. The line shows the ideal result. Reprinted from M. Rana, G. Hamarneh, and J. M. Wakeling, "Automated tracking of muscle fascicle orientation in B-mode ultrasound images," Journal of Biomechanics, vol. 42, no. 13, pp. 2068–2073, 2009, with permission from Elsevier

When the level of noise increased in the synthetic images there was an increase in the error of the estimate of orientation (Fig. 3.20). The errors for both Radon transform and wavelet analysis were less than 0.02° for CNR greater than 0.8. The CNRs for the ultrasound images for the vastus lateralis were in the range of 0.85–1.34.

The ultrasound images showed that the fascicle orientations changed in a cyclical fashion during each pedal cycle. The greatest fascicle angles occurred at the bottom of the pedal cycle when the knee was most extended, with a short vastus lateralis

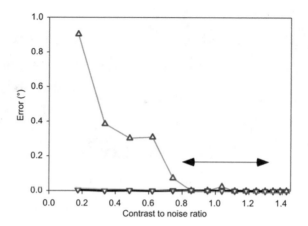

Fig. 3.20 Errors in the estimated orientations from the simulated grids calculated for a range of contrast to noise ratios at a fixed orientation of 8.6°. Angles calculated using wavelet transform are shown by the gray line and triangles with an apex at the top, and angles calculated using Radon transform are shown by the black line and triangles with an apex at their bottom. The arrows show the range of contrast to noise ratios observed in ultrasound images from the vastus lateralis. Reprinted from M. Rana, G. Hamarneh, and J. M. Wakeling, "Automated tracking of muscle fascicle orientation in B-mode ultrasound images," Journal of Biomechanics, vol. 42, no. 13, pp. 2068–2073, 2009, with permission from Elsevier

length. Manual digitization showed that the mean fascicle angles, relative to the x-axis on each image, varied between 2.0° and 3.5° and this corresponded to pennation angles of 8.6–9.5° that were relative to the deep aponeurosis. Considerable variability occurred in the fascicle orientations that were manually digitized by the 10 researchers (Fig. 3.21). The standard deviation for the fascicle orientations for each frame had a mean value of 1.41° (range 0.46–2.83°).

The results from wavelet analysis and Radon transform on the ultrasound images can be visualized in Fig. 3.21. Lines have been drawn that have an orientation determined from the Radon transform or mean wavelet value, and pass through the center of the muscle. In some images wavelet transform resulted in orientations at lesser angles than for Radon transform. The mean orientation α from wavelet analysis was significantly different from the mean fascicle orientation that was manually digitized from each ultrasound frame (two-tailed, matched pair, t-test; p < 0.001) with the mean difference being −1.35°; this difference was less than the standard deviation of the manually digitized values for each frame of 1.41° (Fig. 3.21). There was no significant difference between the dominant orientation θ from Radon transform and the manually digitized values from each ultrasound frame (two-tailed, matched pair, t-test; p = 0.773).

Both the wavelet analysis and Radon transform methods assessed in this study can be applied to an image with no prior manual digitization or algorithm training, and they can be applied to a single frame as easily as to a sequence. These features represent improvements to previous manual digitization and tracking methods [39].

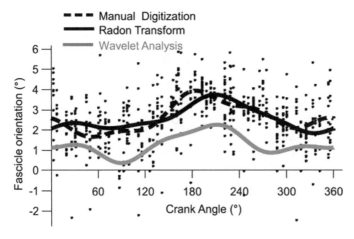

Fig. 3.21 Muscle fascicle orientations in the vastus lateralis during cycling [36]. Orientations are relative to the ultrasound probe (skin) surface. Points show the orientations determined from manual digitization by 10 researchers. Lines show the orientations determined using Fourier series from the manually digitized points (dashed black line), Radon transform (solid black line), and wavelet transform (solid gray line). Reprinted from M. Rana, G. Hamarneh, and J. M. Wakeling, "Automated tracking of muscle fascicle orientation in B-mode ultrasound images," Journal of Biomechanics, vol. 42, no. 13, pp. 2068–2073, 2009, with permission from Elsevier

Radon transform can be used to identify the dominant fascicular orientation within an image, and thus used to estimate muscle fascicle lengths. Wavelet analysis additionally provides information on the local fascicle orientations and can be used to quantify fascicle curvatures and regional differences with fascicle orientation across an image.

More recently, Zhou et al. [28] considered that during muscle contraction, the fascicles may not only change in orientation and length, but also their contrasts in ultrasound images do, and this makes the orientation estimation of individual fascicles not so reliable using the above methods. For a typical ultrasound image of muscle, such as the gastrocnemius muscle as shown in Fig. 3.22, the process of estimating the pennation angle involves the measurement of orientations of the aponeuroses and the fascicle region between aponeuroses. So they proposed a new method that uses RVHT to estimate the aponeurosis orientation and to use the dominant texture orientation of the fascicle region to represent the fascicle orientation. Note that in estimation of the fascicle orientation, the information from a region rather than several individual fascicles is employed. Then the pennation angle is estimated as the difference between orientation of the fascicle and the deep aponeurosis. Since the measurement is automatic, it can be used to measure the pennation angle in each ultrasound image collected during muscle contraction.

Estimation of deep aponeurosis orientation The estimation of deep aponeurosis orientation (O2 in Fig. 3.22) was based on the methods that we have reported earlier,

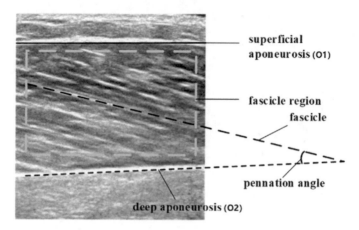

Fig. 3.22 Typical sonogram of the medial gastrocnemius muscle

which include using Gabor filtering to enhance ultrasound images and RVHT to estimate the orientation [21, 40]. This process was repeated for each frame of ultrasound images, and the orientation change was recorded using the RVHT method.

Estimation of dominant fascicle orientation of the selected region To compute the dominant orientation of the selected region of fascicle in ultrasound images, such as the O1 region shown in Fig. 3.22, the local orientation field was acquired in a multi-step process for each pixel. The original region, $I(i, j)$, was first smoothed with a Gaussian filter. Then at each pixel, the gradients, $\partial_x(i, j)$ and $\partial_y(i, j)$, were computed and then the primary local orientation for each pixel was computed using Rao's scheme [20]:

$$Sum_x(i, j) = \sum_{u=i-w/2}^{i+w/2} \sum_{v=j-w/2}^{j+w/2} 2\partial_x(u, v)\partial_y(u, v) \tag{3.5}$$

$$Sum_y(i, j) = \sum_{u=i-w/2}^{i+w/2} \sum_{v=j-w/2}^{j+w/2} \left(\partial_x^2(u, v) - \partial_y^2(u, v)\right) \tag{3.6}$$

$$\theta(i, j) = \frac{\pi}{2} + \frac{1}{2}tan^{-1}\left(\frac{Sum_y(i, j)}{Sum_x(i, j)}\right) \tag{3.7}$$

where $\theta(i, j)$ is the least squares estimation of the orientation at pixel (i, j) and $w \times w$ defines its neighborhood area involved. The reliability coefficient of the orientation field [21] was measured by:

$$R(i, j) = \left(\partial_i^2 + \partial_j^2\right)^{1/2} \frac{\sum_{(k,l)\in\Gamma} \left| \left(\partial_i^2 + \partial_j^2\right)^{1/2} * \cos(\theta(i, j) - \theta(k.l)) \right|}{\sum_{(k,l)\in\Gamma} \left(\partial_k^2 + \partial_l^2\right)^{1/2}} \qquad (3.8)$$

where Γ is a small neighboring region of the pixel (i, j), and its size is related to the local frequency of strongly oriented patterns. The reliability coefficient here is a number between 0 and 1, and its two extremities, 0 and 1, correspond to the isotropic region and the strongly oriented pattern, respectively. To estimate the dominant orientation of the selected region, we used the median value of the orientations for each pixel in the region, as long as its reliability coefficient was larger than an empirically pre-defined threshold, 0.6 in this realization. The fascicle pennation angle was computed as the difference between the dominant orientation of the selected fascicle region O1 and the deep aponeurosis orientation O2 (Fig. 3.22).

Estimation of the dynamic changes of pennation angle In the first image of a series of ultrasound images collected during muscle contraction, the fascicle region of interest was manually selected and its orientation was calculated based on the proposed method. Meanwhile, the orientation and location of the deep aponeurosis of interest were detected, with visual verification. The procedure for processing the 1st frame and calculating the pennation angle is described in Fig. 3.23a. Since the location and orientation of the deep aponeurosis change little during contraction, its change between two frames was confined within a certain range. This helps to track

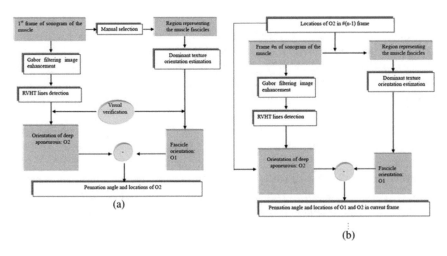

Fig. 3.23 Procedures of estimating the pennation angles. **a** Procedure of estimating the pennation angle in the first frame of the ultrasound image sequence and **b** Procedure of estimating the pennation angle in the subsequent frames following the first frame. Reprinted with permission from Y. Zhou, J.-Z. Li, G. Zhou, and Y.-P. Zheng, "Dynamic measurement of pennation angle of gastrocnemius muscles during contractions based on ultrasound imaging," Biomedical engineering online, vol. 11, no. 1, pp. 1–10, 2012. https://doi.org/10.1186/1475-925X-11-63

consistently the same aponeurosis during muscle contraction. For the selected fascicle region, the dominant orientations at the same region of interest in the subsequent frames were computed using the same procedures as those for the first frame, as shown in Fig. 3.23b. Thus, the dynamic changes of the pennation angle could be automatically estimated frame by frame as the muscle contracts and relaxes.

The method for fascicle orientation estimation was evaluated using synthetic images with different noise levels and later on 500 ultrasound images of human gastrocnemius muscles during isometric plantarflexion. The muscle fascicle orientations were also estimated manually by two operators. From the results it has been found that the proposed automatic method demonstrated a comparable performance to the manual method.

Since the pennation angle may vary in different parts of the image, Sissel deemed that it is necessary to determine from which part of the fascicle plane one should estimate the pennation angle [6]. In addition, one also needs to correct for the angle of the aponeurosis. It was focused on firstly how to calculate the pennation angle without accounting for the deep aponeurosis, and then calculate the angle of the deep aponeurosis to find the final pennation angle.

The pennation angle with respect to the horizontal line The images were measured manually within the lower third of the muscle thickness. In order to make the results as comparable to the manual measurements as possible, the pennation angle should also be estimated within this area. In addition, because the angles might be more uncertain near the edges, the area should not extend all the way to the edges of the image.

To fulfill these requirements, it was chosen to estimate the pennation angle from the angles measured in a window consisting of all points located in the middle third in the horizontal direction, and the lower third of the fascicle plane in the vertical direction. In addition, the lower border of the window was moved 1/12 of the fascicle plane upwards, while the upper border of the window was moved by the same amount downwards. The last step was performed since the angles were calculated in an area of 30×70 around the points inside the window.

The image was filtered with the vesselness filter, and each measurement was calculated in a 30×70 window around the points in question, and normalized Radon transform was performed to locate an angle in each 30×70 window. That procedure provided about 100 measurements from which to determine the pennation angle. The median was used to find the peak value in the area, and the resulting value was used as the pennation angle. However, this pennation angle is only valid if the deep aponeurosis is completely horizontal. It rarely is, so one needs to correct the pennation angle with respect to the angle of the aponeurosis.

The pennation angle with respect to the deep aponeurosis To determine the angle of the aponeurosis with regard to the pennation angle, two approaches were used. One approach was to simply fit a line to the quadratic curve, only using the middle 90% of the data, and calculate the angle of this line compared to the horizontal line. The second approach was to only fit a line to the part of the quadratic curve below the window defined above. Both these approaches seemed to give plausible results, and in many images there was little difference between the two variations. Finally, the pennation angle was found by calculating the angle between the line determined from the window, and the line fitted to the deep aponeurosis.

The performance of the method was tested on two different datasets, and produced mostly good results when the aponeuroses were detected correctly. The results proved that when finding the pennation angle, it was important to detect the dominant orientation of the muscle fascicles accurately, as well as robust detection and good representation of the aponeuroses.

More recently, Yuan et al. [41] suggested that none of the above-mentioned methods solve the existing problems sufficiently in the dynamic measurements of pennation angle, including (1) the measurement errors of pennation angle are not satisfactory, (2) the results of pennation angle are jittered, while the contractions of muscles are smooth apparently. Since skeletal muscle is one of the most active tissues in the human body its dynamic change should be a regular progression in the process of contraction of the muscle system [42]. Therefore, they proposed a weighted-average (based on the pennation angle from a vast amount of muscle fascicles) method based on gradient Radon transform to realize a fully automatic and precise measurement of pennation angle during muscle contraction in ultrasound images. The procedures of the proposed methods include (1) automatic apportionment of the muscle structure and (2) measurement of the pennation angle.

The automatic apportionment of the muscle structure Standard Radon transform represents the grayscale integral of the entire image plane in its projection orientation. Because the peak points on the Radon transform matrix can characterize the straight line, Radon transform is widely used in the extraction of lines and the recognition of boundaries. However, the edge effect occurs quickly due to different lengths of the integral path in different projection orientations. As shown in Fig. 3.24a, it is impossible to determine which peak points represent the orientation of the aponeuroses. To eliminate the edge effect and determine the aponeuroses boundary, gradient Radon transform has been proposed, which is defined as follows:

$$\frac{\partial}{\partial \rho} GR(\rho, \theta) = \iint_D (I(x, y) - mean(I))\delta(\rho - x\cos\theta - y\sin\theta)dxdy \qquad (3.9)$$

where $\frac{\partial}{\partial \rho} GR(\rho, \theta)$ represents the gradient Radon transform matrix, and $mean(I)$ represents the mean gray value of the original image. As shown in Fig. 3.24b, c, when one image was applied by Eq. (3.3), the gradient Radon transform matrix $G1$ can be obtained to locate the lower edge of the muscle structure. In the same way, $G2$ was obtained to locate the upper edge using the reversing phase (Fig. 3.24d). To accurately locate aponeuroses and divide the muscle structure, the following steps need to be completed, and the detailed process using an example is shown in Fig. 3.25.

• Extract vast peak points in $G1$ and remove highlight noises caused by image edges.

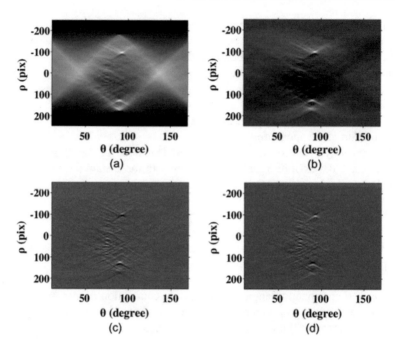

Fig. 3.24 Radon transform matrix type. **a** Standard Radon transform matrix. **b** The result of subtracting the gray mean value from the original image in the Radon transform matrix. **c** Gradient Radon transform matrix. **d** Gradient Radon transform matrix reversing phase. Reprinted from C. Yuan, Z. Chen, M. Wang, J. Zhang, K. Sun, and Y. Zhou, "Dynamic measurement of pennation angle of gastrocnemius muscles obtained from ultrasound images based on gradient Radon transform," Biomedical Signal Processing and Control, vol. 55, p. 101,604, 2020, with permission from Elsevier

- Use a Gaussian mixture model for cluster analysis. The peak points were divided into the regions of superficial aponeurosis, deep aponeurosis, and fascicles.
- Remove the peak points with gray values below average in the regions of the superficial or deep aponeurosis.
- Repeat the above steps in $G2$.

It is well-known that the gray value on the gradient Radon transform matrix represents the cumulative result of mapping each straight outline on the original image from Euclidean space to Radon space. The larger the gray value, the higher the possibility that the inverse transformation of the feature point back to Euclidean space becomes a straight line. In this study, the top 0.1% of the peak points were extracted, and their respective gray values were used as weights to weigh their orientation and position. Since there is no difference between the angles of the upper and lower edges of the deep aponeurosis, the weighted average angle between the lower edge of the deep aponeurosis and the horizontal line was selected and recorded as θ_1. The weighted average is calculated as follows:

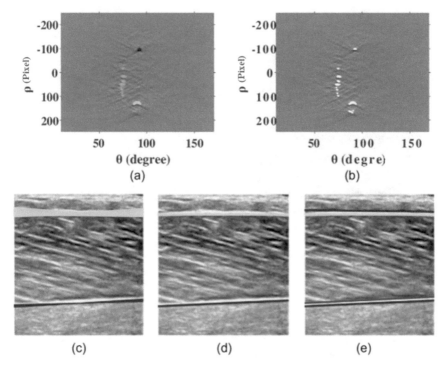

Fig. 3.25 The apportionment of ultrasound images of skeletal muscle. **a** Extraction of vast peak points in *G1*. **b** Cluster analysis in *G1*. Green, yellow, and purple points represent the groups of fascicles and the lower edges of the superficial and deep aponeurosis, respectively. **c** Mapping of the peak points between groups of lower edges of the superficial and deep aponeurosis in *G1* to form lines. **d** The results of weighted average in *G1*. **e** Location of aponeuroses through *G1* and *G2*. Red, yellow, blue, and purple lines represent the upper and lower edges of aponeuroses respectively. Reprinted from C. Yuan, Z. Chen, M. Wang, J. Zhang, K. Sun, and Y. Zhou, "Dynamic measurement of pennation angle of gastrocnemius muscles obtained from ultrasound images based on gradient Radon transform," Biomedical Signal Processing and Control, vol. 55, p. 101,604, 2020, with permission from Elsevier

$$\tilde{\theta} = \frac{\sum \theta_i \times w_i}{\sum w_i} \tag{3.10}$$

where $\tilde{\theta}$ represents the weighted average orientation of all the fascicles, θ_i represents the orientation of each bundle of fascicles, and w_i represents the weight of each bundle of fascicles. This study shows that the proposed method can accurately identify the dominant orientation and position of aponeuroses on the muscle from the ultrasound image, thereby realizing the automatic division of the muscle structure.

Measurement of the pennation angle Through previous operations, the vast peak points representing the lower edge of the muscle structure in *G1* and the upper edge of the muscle structure in *G2* respectively, were obtained. Then, the peak points

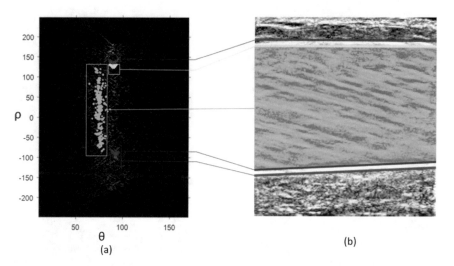

Fig. 3.26 Calculation of the pennation angle. **a** peak points in the Gradient Radon transform matrix. **b** The labeled (with different colors) structures in the ultrasound image corresponding to peaks in (a)

representing the fascicles between *G1* and *G2* were combined to form vast fascicles in the Euclidean space. As shown in Fig. 3.26a, the average orientation and position of all the fascicles were weighted, and the angle θ_2 between the vast-fascicle-weighted average orientation and the horizontal line was recorded. With the above process, the muscle pennation angle of each frame image can be accurately measured (Fig. 3.26b).

Also, this study aimed to realize a fully automatic and precise measurement of the pennation angle during muscle contraction in ultrasound images. The coefficient of multiple correlations (CMC) and the linear regression with a Bland–Altman analysis were implemented to evaluate the reliability. Pearson correlation analysis and polynomial regression analysis were applied to describe the association between the pennation angle and the corresponding muscle torque. Meanwhile, a new criterion, roughness, has been defined to assess the smoothness of the time-pennation angle curve during muscle contraction. The experimental results indicate that the time-angle curves calculated by the proposed method are highly correlated with the torque of muscle contraction (Pearson correlation coefficient = 0.91 ± 0.04, R = 0.87 ± 0.14), in closer agreement with manual results (CMC = 0.83 ± 0.15) and smoother (RN = 0.26 ± 0.10) than other state-of-the-art methods. In conclusion, the proposed novel method can solve the limitations as reported in the earlier methods in the calculation of pennation angle, and subsequently provide a promising way for functional muscle assessment or prosthesis control.

3.5   Pre-processing Techniques

The performance of the method such as RVHT, is closely related to the image quality, *i.e.*, noisier sonograms lead to poorer performance of fiber detection and angle estimation compared with manual drawing and reading by operators. To overcome this problem, as introduced in Sect. 2.4.2, pre-processing techniques are often used to enhance the ultrasound images. The popular methods include MVEF, Gabor filtering, etc. For more details, please refer to previous sections such as Sect. 2.4.2.

3.6   Summary

The measurement of skeletal muscle pennation angle using ultrasound imaging has been demonstrated to be feasible, not only manually but automatically using different approaches. Since PA and its change during contraction is closely related to muscle function and strength, it has wide application. It is important to note that ultrasound images present the PA for the imaged longitudinal cross-section of muscle with a certain width controlled by the ultrasound beam. During muscle contraction, particularly during dynamic motion of the subject, the relative alignment between ultrasound probe and muscle changes; thus, the ultrasound images may be changed by the body posture as well. This needs to be carefully noted during any study using pennation angle for dynamic muscle analysis. Additional research works are also needed in the future to tackle this issue, such as using a 2D array transducer to obtain real-time 3D imaging for muscle to analyze the PA in 3D dynamically. Another important direction for future research and development is the attachment of ultrasound probes during dynamic motion of the subject. A wireless wearable ultrasound scanner which can be attached to the skin surface is an ideal solution to tackle this issue, and this is another future direction for ultrasound imaging for muscle. A recent study has already demonstrated such feasibility [43].

References

1. Zatsiorsky, V.M., Prilutsky, B.I.: Biomechanics of skeletal muscles. Biomechanics of Skeletal Muscles (2012)
2. Fukunaga, T., Ichinose, Y., Ito, M., Kawakami, Y.S., Fukashiro, S.: Determination of fascicle length and pennation in a contracting human muscle in vivo. J. Appl. Physiol. **82**(1), 354 (1997)
3. Narici, M.: Human skeletal muscle architecture studied in vivo by non-invasive imaging techniques: functional significance and applications. J. Electromyogr. Kinesiol. **9**(2), 97–103 (1999)
4. Caresio, C., Salvi, M., Molinari, F., Meiburger, K.M., Minetto, M.A.: Fully automated muscle ultrasound analysis (MUSA): robust and accurate muscle thickness measurement. Ultrasound Med. Biol. **43**(1), 195–205 (2017)

5. Chen, X., Li, Q., Qi, S., Zhang, H., Chen, S., Wang, T.: Continuous fascicle orientation measurement of medial gastrocnemius muscle in ultrasonography using frequency domain Radon transform. Biomed. Sig. Process. Control **20**, 117–124 (2015)
6. Fladby, S.: Automatic feature extraction and parameter estimation from unipennate muscles in B-mode ultrasound images (2017)
7. Maganaris, C.N., Baltzopoulos, V., Sargeant, A.J.: Repeated contractions alter the geometry of human skeletal muscle. J. Appl. Physiol. **93**(6), 2089–2094 (2002)
8. Wickiewicz, T.L., Roy, R.R., Powell, P.L., Edgerton, V.R.: Muscle architecture of the human lower limb. Clin. Orthop. Relat. Res. **179**, 275–283 (1983)
9. Friederich, J.A., Brand, R.A.: Muscle fiber architecture in the human lower limb. J. Biomech. **23**(1), 91–95 (1990)
10. Sacks, R.D., Roy, R.R.: Architecture of the hind limb muscles of cats: functional significance. J. Morphol. **173**(2), 185–195 (1982)
11. Gans, C.: The functional significance of muscle architecture: a theoretical analysis. Adv. Anat. Embryol. Cell Biol. **38**, 115–142 (1965)
12. Zhou, G.-Q., Zheng, Y.-P.: Automatic fascicle length estimation on muscle ultrasound images with an orientation-sensitive segmentation. IEEE Trans. Biomed. Eng. **62**(12), 2828–2836 (2015)
13. Zhou, G.-Q., et al.: Automatic myotendinous junction tracking in ultrasound images with phase-based segmentation. BioMed Res. Int. **2018** (2018)
14. Rekabizaheh, M., Rezasoltani, A., Lahouti, B., Namavarian, N.: Pennation angle and fascicle length of human skeletal muscles to predict the strength of an individual muscle using real-time ultrasonography: a review of literature. J. Clin. Physiotherapy Res. **1**(2), 42–48 (2016)
15. Hough, P.V.: Method and means for recognizing complex patterns. ed: Google Patents (1962)
16. Immerkær, J.: Some remarks on the straight line Hough transform. Pattern Recogn. Lett. **19**(12), 1133–1135 (1998)
17. Hart, P.E., Duda, R.: Use of the Hough transformation to detect lines and curves in pictures. Commun. ACM **15**(1), 11–15 (1972)
18. Ballard, D.H.: Generalizing the Hough transform to detect arbitrary shapes. Pattern Recogn. **13**(2), 111–122 (1981)
19. Princen, J., Illingworth, J., Kittler, J.: A hierarchical approach to line extraction based on the Hough transform. Comput. Vis. Graph. Image Process. **52**(1), 57–77 (1990)
20. Song, J., Lyu, M.R.: A Hough transform based line recognition method utilizing both parameter space and image space. Pattern Recogn. **38**(4), 539–552 (2005)
21. Zhou, Y., Zheng, Y.-P.: Estimation of muscle fiber orientation in ultrasound images using revoting hough transform (RVHT). Ultrasound Med. Biol. **34**(9), 1474–1481 (2008)
22. Kiryati, N., Bruckstein, A.M.: Antialiasing the Hough transform. CVGIP Graph. Models Image Process. **53**(3), 213–222 (1991)
23. Reeves, N.D., Narici, M.V., Maganaris, C.N.: In vivo human muscle structure and function: adaptations to resistance training in old age. Exp. Physiol. **89**(6), 675–689 (2004)
24. Trouvé, E., Mauris, G., Rudant, J.-P., Tonyé, E.: Detection of linear features in synthetic-aperture radar images by use of the localized Radon transform and prior information. Appl. Opt. **43**(2), 264–273 (2004)
25. Copeland, A.C., Ravichandran, G., Trivedi, M.M.: Localized Radon transform-based detection of ship wakes in SAR images. IEEE Trans. Geosci. Remote Sens. **33**(1), 35–45 (1995)
26. Jin, Y., Wang, S.: An algorithm for ship wake detection from the synthetic aperture radar images using the Radon transform and morphological image processing. Imaging Sci. J. **48**(4), 159–163 (2000)
27. Murphy, L.M.: Linear feature detection and enhancement in noisy images via the Radon transform. Pattern Recogn. Lett. **4**(4), 279–284 (1986)
28. Zhou, Y., Li, J.-Z., Zhou, G., Zheng, Y.-P.: Dynamic measurement of pennation angle of gastrocnemius muscles during contractions based on ultrasound imaging. Biomed. Eng. Online **11**(1), 1–10 (2012)

29. Radon, J.: On the determination of functions from their integral values along certain manifolds. IEEE Trans. Med. Imaging **5**(4), 170–176 (1986)
30. Aggarwal, N., Karl, W.C.: Line detection in images through regularized Hough transform. IEEE Trans. Image Process. **15**(3), 582–591 (2006)
31. Jensen, J.A., Svendsen, N.B.: Calculation of pressure fields from arbitrarily shaped, apodized, and excited ultrasound transducers. IEEE Trans. Ultrason. Ferroelectr. Freq. Control **39**(2), 262–267 (1992)
32. Rehorn, M.R., Blemker, S.S.: The effects of aponeurosis geometry on strain injury susceptibility explored with a 3D muscle model. J. Biomech. **43**(13), 2574–2581 (2010)
33. Kararizou, E., Manta, P., Kalfakis, N., Vassilopoulos, D.: Age-related morphometric characteristics of human skeletal muscle in male subjects. Pol. J. Pathol. **60**(4), 186–188 (2009)
34. Zhao, H., Zhang, L.-Q.: Automatic tracking of muscle fascicles in ultrasound images using localized radon transform. IEEE Trans. Biomed. Eng. **58**(7), 2094–2101 (2011)
35. Zhao, H., Ren, Y., Wu, Y.-N., Liu, S.Q., Zhang, L.-Q.: Ultrasonic evaluations of Achilles tendon mechanical properties poststroke. J. Appl. Physiol. **106**(3), 843–849 (2009)
36. Rana, M., Hamarneh, G., Wakeling, J.M.: Automated tracking of muscle fascicle orientation in B-mode ultrasound images. J. Biomech. **42**(13), 2068–2073 (2009)
37. Frangi, A.F., Niessen, W.J., Vincken, K.L., Viergever, M.A.: Multiscale vessel enhancement filtering. In: International Conference on Medical Image Computing and Computer-Assisted Intervention, pp. 130–137. Springer (1998)
38. Jafari-Khouzani, K., Soltanian-Zadeh, H.: Radon transform orientation estimation for rotation invariant texture analysis. IEEE Trans. Pattern Anal. Mach. Intell. **27**(6), 1004–1008 (2005)
39. Loram, I.D., Maganaris, C.N., Lakie, M.: Use of ultrasound to make noninvasive in vivo measurement of continuous changes in human muscle contractile length. J. Appl. Physiol. **100**(4), 1311–1323 (2006)
40. Zhou, Y., Zheng, Y.-P.: Longitudinal enhancement of the hyperechoic regions in ultrasonography of muscles using a gabor filter bank approach: a preparation for semi-automatic muscle fiber orientation estimation. Ultrasound Med. Biol. **37**(4), 665–673 (2011)
41. Yuan, C., Chen, Z., Wang, M., Zhang, J., Sun, K., Zhou, Y.: Dynamic measurement of pennation angle of gastrocnemius muscles obtained from ultrasound images based on gradient Radon transform. Biomed. Sig. Process. Control **55**, 101604 (2020)
42. Frontera, W.R., Ochala, J.: Skeletal muscle: a brief review of structure and function. Calcif. Tissue Int. **96**(3), 183–195 (2015)
43. Ma, C.Z.H., et al.: Towards wearable comprehensive capture and analysis of skeletal muscle activity during human locomotion. Sensors (2019)

Chapter 4
Measurement of Skeletal Muscle Fascicle Length

Abstract Muscle fascicle length (FL) is also an important quantitative indicator of skeletal muscle dynamics, and its changes directly reflect muscle activity. The calculation methods for muscle fiber length include the use of extended field-of-view ultrasonography (EFOV US) technology and optical flow tracking methods, etc. These methods have their own limitations, especially the three-dimensional structure nature and non-absolute-linear nature of muscle fascicles, which pose challenges to the calculation of muscle fiber length.

4.1 Why is Fascicle Length so Important to Measurement?

As shown in Fig. 4.1, muscle length L_m is defined as "the distance from the origin of the most proximal muscle fibers to the insertion of the most distal fibers" [1]. This is not the same as the muscle fiber length (L_f) because of the variable degree of "stagger" seen in muscle fibers as they arise from and insert onto tendon plates (Fig. 4.1). To date, it has only been possible to determine muscle fiber length by microdissection of individual fibers from fixed tissues or by laborious identification of fibers by glycogen depletion on serial sections along the length of the muscle [2]. Unless investigators are explicit, when they refer to muscle fiber length, they are probably referring to muscle fiber bundle length, because it is extremely difficult to isolate intact individual fibers, which run from origin to insertion, especially in mammalian tissues [2, 3]. Thus, when microdissection is performed, bundles consisting of 5–50 fibers are typically used to estimate L_f. Because of the many different methods available for L_f determination, a distinction between "anatomical" fiber length and "functional" fiber length may be made. The anatomical fiber length determined by microdissection may not accurately represent the function of the fibers within the muscle, because fibers may be broken with their activation pattern unknown [4].

Muscle fiber length is an important determinant of the force-generating capability of muscle, with shorter muscle fibers having a narrower force–length relationship, a reduced maximum shortening speed and a reduced length at which they develop passive forces [4]. Lichtwark et al. demonstrated that different muscle fascicle lengths and tendon compliance combinations are required to maximize efficiency under

© Springer Nature Singapore Pte Ltd. 2021   79
Y. Zhou and Y.-P. Zheng, *Sonomyography*, Series in BioEngineering,
https://doi.org/10.1007/978-981-16-7140-1_4

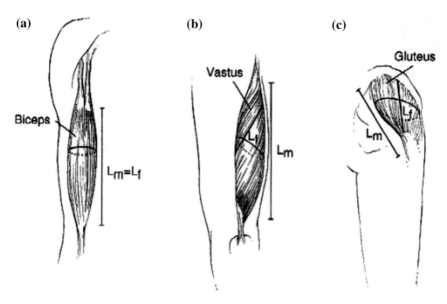

Fig. 4.1 Artist's conception of three general types of skeletal muscle architecture. (A) Longitudinal architecture in which muscle fibers run parallel to the muscle's force-generating axis, as in the biceps brachii. (B) Unipennate architecture in which muscle fibers run at a fixed angle relative to the muscle's force-generating axis, as in the vastus lateralis muscle. (C) Multipennate architecture in which muscle fibers run at several angles relative to the muscle's force-generating axis (gluteus medius muscle). L_f, muscle fiber length; L_m, muscle length. Reprinted from R. L. Lieber, "Skeletal muscle structure and function," Implications for Rehabilitation and Sports Medicine, 1992, with permission from Wolters Kluwer

different gait conditions and speeds, because a different muscle fascicle length will compensate for a stiffer or more compliant tendon to maintain high efficiency [5]. Abe et al. concluded that longer fascicle length is associated with greater sprinting performance in sprinters, while there are no gender differences in fascicle length for elite sprinters [6]. To examine the dynamic function of muscles during contractions and movement, it is necessary to examine the length changes of the muscle fibers themselves.

Magnetic resonance imaging (MRI) is the modality of choice for direct measurement of muscle length in vivo [7, 8]. However, this technique is expensive, not easily accessible, relatively slow for each scan, and the scanning must be done statically with the subject in a supine, prone, or side-lying position, limiting the dynamic measurement of muscle fascicle lengths in other body positions or activities. B-mode ultrasound (US) is commonly used to visualize muscle and tendon morphology and obtain quantitative information concerning muscle properties including muscle anatomical cross-sectional area, muscle thickness, fascicle length, and fascicle angle [9–12]. Use of flat ultrasound probes has made observation of muscle fascicle lengths during activities such as walking and running more feasible.

4.2   Indirect Method (Calculated from Muscle Thickness and Pennation Angle)

Measurement of in vivo muscle length is difficult with B-mode US as the field of view is typically insufficient to visualize the whole muscle. An alternative approach is to use trigonometric estimations (linear extrapolations) to calculate the length of the part of the fascicle that cannot be imaged directly due to the limited field of view of static US imaging [13, 14].

Assuming that aponeuroses and fascicles follow linear paths and aponeuroses are parallel (Fig. 4.1), the extrapolate method calculates the remaining portion of the muscle fascicle by dividing the remaining muscle thickness (h) by the sine of the pennation angle (α).

$$L_f = \frac{h}{\sin \alpha} \tag{4.1}$$

The most common trigonometric estimation to estimate FL [14] is shown in Fig. 4.2. L_l is the measured fascicle length visible in the image area, h is the vertical distance of the fascicle end point from the superficial aponeurosis, and β is the angle between the fascicle and the deep aponeurosis. When the aponeuroses are not parallel, the angle between them is subtracted from the measured pennation angle to make the calculation possible.

However, trigonometric estimation of L_f assumes that the nonimaged fascicle portion follows a linear path, which is rarely true in skeletal muscle [13]. As shown in Fig. 4.3, some curvature exists in almost every visible fascicle, and especially at the distal ends of the fascicles close to the superficial and deep aponeuroses. This curvature also makes it difficult to accurately measure the pennation angle, which is a component of L_f trigonometric extrapolation. When the actual shape of the fascicle is curved, the result of trigonometric estimation is inaccurate [15] (Fig. 4.4).

Fig. 4.2 When aponeuroses and fascicle follow linear paths and aponeuroses are parallel, the fascicle length can be easily estimated from geometric relationships

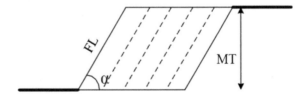

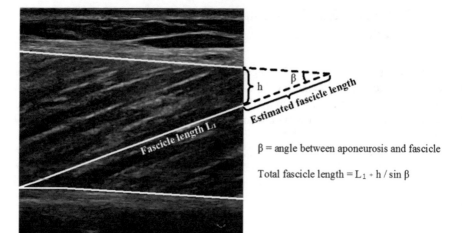

β = angle between aponeurosis and fascicle

Total fascicle length $= L_1 + h / \sin \beta$

Fig. 4.3 Total fascicle length was estimated using a linear continuation of aponeurosis and fascicles when the fascicle was not fully visible within the imaged area. Reprinted with permission from T. Finni, S. Ikegawa, V. Lepola, and P. Komi, "Comparison of force–velocity relationships of vastus lateralis muscle in isokinetic and in stretch-shortening cycle exercises," Acta Physiologica Scandinavica, vol. 177, no. 4, pp. 483–491, 2003

Fig. 4.4 Directions of superficial aponeurosis, deep aponeurosis, and fascicle are extended by straight lines (solid black lines). Dotted lines, unpredictable curvature of fascicle

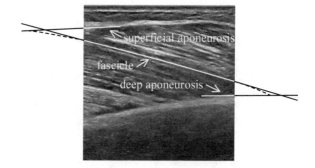

4.3   Direct Method

4.3.1   FL Computation Using Extended Field-Of-View Ultrasonography

The extended field-of-view ultrasonography (EFOV US) technique [16, 17], which uses an algorithm to automatically fit series of images, allows scanning of entire fascicles within one continuous scan [18, 19] and observation of long muscle fascicles under static conditions (Fig. 4.5) [15]. The programmed reconstruction algorithm recognizes the overlapping regions within the real-time images obtained in series when the probe is moved along the imaged surface. In generating the EFOV image in

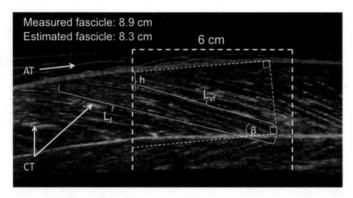

Fig. 4.5 Example vastus intermedius muscle (VL) extended field-of-view ultrasonography (EFOV US) image. The white dashed line represents the 6-cm-wide single scan used for comparison; the black continuous line is the analyzed fascicle (L_f), L_{vf} is the visible part of the fascicle on the 6-cm scan, β is the angle between the fascicle and a line in parallel with the superficial aponeurosis, and h is the vertical distance from the superficial aponeurosis to the crossing point with the fascicle. Adipose and connective tissue markings, as shown, were used to ascertain that the imaging plane was similar in repeated scans. In this image, the fascicle length difference between directly measured and estimated fascicle lengths was 4.8%. Reprinted with permission from M. Noorkoiv, A. Stavnsbo, P. Aagaard, and A. J. Blazevich, "In vivo assessment of muscle fascicle length by extended field-of-view ultrasonography," Journal of Applied Physiology, vol. 109, no. 6, pp. 1974–1979, 2010

real-time, the pattern-matching technology is applied to the region of interest (ROI), which is fixed in the system. Matching of image features in successive frames is accomplished by detecting the brightness level in a B-mode image and then searching the prior frame (i.e., the reference frame) for regions in the ROI with similar properties. The different parts of the detected images are subsequently attached to the previous image. This allows determination of overall probe motion by detecting the movement direction [15].

However, it is also the case for conventional static US imaging. L_f can be distorted by the orientation of the ultrasound probe [20, 21], so the probe must be aligned with the orientation of the fascicles to minimize perspective and parallax measurement errors. Furthermore, misalignment of the ultrasound probe in relation to the fascicular plane would result in overestimation of fascicle length [22] and underestimation of fascicle angle [22, 23]. This might be especially problematic when using the EFOV US technique since the probe has to be moved over a longer distance within one continuous scan (Fig. 4.6). Also the probe must be kept perpendicular to the fascicular plane (i.e., when the fascicles are best visible) along the whole imaging path, which is especially problematic in muscles with curved deep aponeuroses [23]. Another limitation of the EFOV US technique is that it is not applicable during dynamic muscle contractions and image acquisition may take too long to be useable during maximal isometric contraction.

The EFOV US technique, when done by experienced researchers after extensive practice, is a reliable and valid method with which to measure fascicle length in vivo

Fig. 4.6 Measurement of vastus intermedius muscle (VL) fascicle length (L_f). The black line marked on the skin is the line of orientation of VL fascicles, imaged by ultrasound. Reprinted with permission from M. Noorkoiv, A. Stavnsbo, P. Aagaard, and A. J. Blazevich, "In vivo assessment of muscle fascicle length by extended field-of-view ultrasonography," Journal of Applied Physiology, vol. 109, no. 6, pp. 1974–1979, 2010

in relaxed muscles when fascicles are too long to be visualized by conventional static ultrasound scanning [15].

4.3.2 FL Computation Using Optical Flow

In 2011, Cronin et al. [24] implemented an efficient automatic fascicle tracking method based on the Lucas-Kanade optical flow algorithm [25] with an affine optic flow extension [26]. Using this algorithm, it is possible to determine the global movement of visible muscle from one frame to the next during muscle contraction. This type of tracking is physiologically relevant because muscle fascicles both shorten and rotate during muscle contraction, which can be modeled by an affine transformation.

In the first frame for analysis, the examiner manually selects the muscle region of interest and defines the average initial length of the muscle fascicles by defining fascicle end points. The region of interest of muscle is defined as the area between the superficial and deep fascia of the MG muscle that is visible in the ultrasound image. Muscle fascicle length is defined as the straight line distance between the superficial and the deep muscular fascia parallel to the lines of collagenous tissue visible in the ultrasound image (Fig. 4.7). All measurements are made in the middle of the image where the full length of the fascicle can be visualized in the first ultrasound frame. Fascicle length changes are subsequently tracked using a Lucas-Kanade optical flow algorithm with affine optic flow extension. The algorithm determines the optical flow (spatial and temporal gray-level gradients) between two consecutive images within a defined region of interest and uses a least-squares fit of the affine transformations (translation, dilation, rotation and shear) to best represent that flow pattern. The

Fig. 4.7 2D B-mode
ultrasound image showing
the region of interest, defined
fascicle length, and
superficial and deep
aponeuroses. Collagenous
and connective tissue appears
as white in the image, and
the interpreted line of action
of the fascicle length
between the superficial and
deep aponeurosis appears as
the dark space

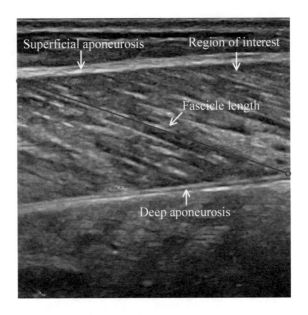

region of interest is the area of muscle to be tracked, i.e. the area inside the muscle
fascia, which is defined in the initial frame using a polygon shape drawn within the
muscle borders (Fig. 4.7). The muscle must move in a uniform manner between
frames that fit the affine transformation.

The affine optic flow model has six parameters as shown in Eq. 4.2: vxt, optic
flow at origin (top left corner) in the x-direction; vyt, optic flow at origin (top left
corner) in the y-direction; d, rate of dilation; r, rate of rotation; $s1$, shear along the
main image axis; $s2$, shear along the diagonal axis.

$$(vx, vy) = [xt\,1] \times \begin{bmatrix} d + s1 & s2 + r \\ s2 - r & d - s1 \\ vxt & vyt \end{bmatrix} \tag{4.2}$$

The calculated affine transformation is applied to the defined fascicle end points
from one frame to the next. This iterative approach allows fascicle length to be
defined for each ultrasound image in a sequence (see Fig. 4.8 for a schematic of the
workflow for each iteration).

Gillett et al. [27] have determined the within- and between-examiner reliability
of in vivo MG fascicle length changes from 2D B-mode ultrasound using the affine
flow algorithm. The affine optical flow algorithm is a promising method for tracking
muscle deformation during contractions. One advantage of the algorithm is that it
provides a high level of tracking stability because it tracks global movement over
a large region of interest (i.e. the visible muscle). Therefore, the algorithm is less
susceptible to drift over time due to the accumulation of small errors in individual
frame-to-frame tracking when compared with tracking small regions of interest. A

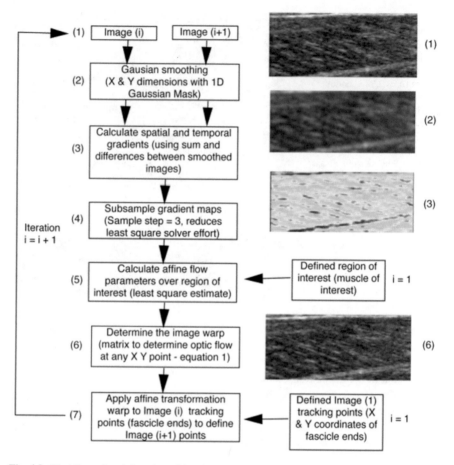

Fig. 4.8 Workflow of each iteration of the algorithm (frame by frame) used in the present study to track muscle fascicle endpoints. The diagram includes a graphical representation of the processing involved in each step with numbers to the right of the images corresponding to each step in the workflow. 1, original image; 2, smoothed image; 3, spatial gradient in the y-direction, represented as a heat map; 6, calculated flow at different points in images as represented by red vector arrows. Reprinted with permission from N. J. Cronin, C. P. Carty, R. S. Barrett, and G. Lichtwark, "Automatic tracking of medial gastrocnemius fascicle length during human locomotion," Journal of Applied Physiology, vol. 111, no. 5, pp. 1491–1496, 2011

further advantage of this algorithm is its ability to track the movement of the muscle fascicle even if the defined end points of the fascicle are outside the region of interest, meaning that features can be virtually tracked outside the image. This is possible because affine transformation can be applied to any point in 2D space.

4.4 Applications and Relationship to Muscle Elasticity

With regard to the estimation of muscle force, recent studies using an ultrasound-based elastography technique named supersonic shear imaging (SSI) have reported that muscle shear elastic modulus (or shear modulus, a measure of normalized stiffness) can reliably and accurately predict passive [28] and active [29, 30] forces exerted by individual muscles. The modulus is calculated as a function of the propagation velocity of shear waves induced by a focused ultrasound beam [31], along the axis of whole muscle contraction. Considering a close link between active force and stiffness in isolated muscle fibers [32], it is likely that the length-dependent changes in muscle force production also manifest as changes in muscle shear modulus.

Sasaki et al. [33] reported that the shear modulus of the tetanized human tibialis anterior muscle measured along the axis of contraction increased with increases of both fascicle length and contractile force (Fig. 4.9). The results also indicated the linear association between muscle force and shear modulus, providing novel evidence that length–force relationship, one of the most fundamental characteristics of muscle, can be inferred from in vivo imaging of shear modulus in the tibialis anterior muscle. Furthermore, a similar length dependence of muscle shear modulus was observed during maximal voluntary contractions in which neural and mechanical interactions of multiple muscles are involved.

This study is the first of its kind to determine muscle shear modulus in vivo during both tetanic and voluntary isometric contractions using a relatively new imaging technique (SSI) (Fig. 4.9) and, further, to explore the association of shear modulus

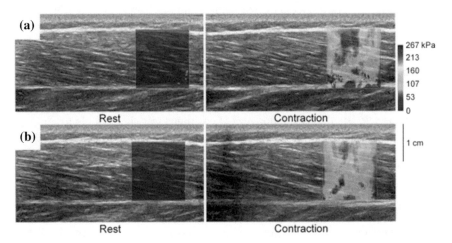

Fig. 4.9 Typical color-coded shear modulus distribution superimposed on a longitudinal ultrasound image of the tibialis anterior muscle at rest and during contraction. A: images obtained in tetanic contraction (TC) session. B: images obtained in MVC session. Note that the shear modulus substantially increased during contraction compared with at rest. Reprinted with permission from K. Sasaki, S. Toyama, and N. Ishii, "Length–force characteristics of in vivo human muscle reflected by supersonic shear imaging," Journal of Applied Physiology, vol. 117, no. 2, pp. 153–162, 2014

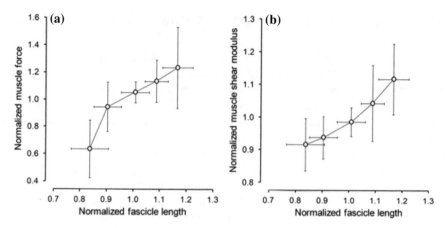

Fig. 4.10 Associations of muscle force (A) and shear modulus (B) with fascicle length of the tetanized tibialis anterior muscle. Data are normalized to the average of five different joint positions in each participant and expressed as means and SD (n = 9). Regression analysis revealed significant positive associations of muscle force ($R^2 = 0.51$, n = 45, p < 0.001) and shear modulus ($R^2 = 0.42$, n = 45, p < 0.001) with fascicle length. Reprinted with permission from K. Sasaki, S. Toyama, and N. Ishii, "Length–force characteristics of in vivo human muscle reflected by supersonic shear imaging," Journal of Applied Physiology, vol. 117, no. 2, pp. 153–162, 2014

with muscle fascicle length and contractile force. In connection with the length–force relationship of muscle, the length-shear modulus relationship provides not only valuable information on the capacity of SSI to estimate individual muscle force, but also a clue to understanding the structures and mechanisms responsible for the stiffness of an active human muscle. The results demonstrated the length dependence of muscle shear modulus, which was similar to that of contractile force (Fig. 4.10).

In the MVC session, the muscle shear modulus increased with fascicle length in a manner similar to that observed in the TC session (Fig. 4.11). This result suggests that SSI can assess the muscle shear modulus and its length dependence, irrespective of contraction mode. It should be noted, however, that the shear modulus measured at short fascicle lengths (i.e., at dorsiflexed positions) was significantly lower in the MVC session than in the TC session. This observation may not impair the validity of SSI to estimate individual muscle force during voluntary contractions, but rather reflect the actual difference in the contractile force between tetanic and voluntary contractions.

To summarize, as an important and direct indicator of muscle activity, FL has attracted great attention, yet its automatic calculation methods presented here still have their own limitations and great space to improve. For example, there are challenges to handling the nature of the three-dimensional structure of muscles and the feature of non-absolute linearity of muscle fascicles. Further research works in these areas are in demand for wider application of fascicle length.

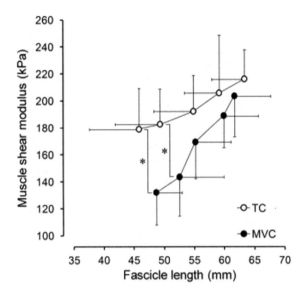

Fig. 4.11 Comparison of length-shear modulus characteristics of the tibialis anterior muscle determined in TC session (○) and MVC session (●). Data are means and SD (n = 9). Two-way repeated-measures ANOVA revealed a significant interaction of "session" and "joint position" on the muscle shear modulus (p = 0.045), but not on the fascicle length (p = 0.20). *Significant difference between the two sessions (p < 0.05, Student's paired t-test with the false discovery rate procedure). Reprinted with permission from K. Sasaki, S. Toyama, and N. Ishii, "Length–force characteristics of in vivo human muscle reflected by supersonic shear imaging," Journal of Applied Physiology, vol. 117, no. 2, pp. 153–162, 2014

References

1. Lieber, R.L.: Skeletal muscle structure and function. Inplications for rehabilitation and sports medicine (1992)
2. Ounjian, M., et al.: Physiological and developmental implications of motor unit anatomy. J. Neurobiol. **22**(5), 547–559 (1991)
3. Loeb, G., Pratt, C., Chanaud, C., Richmond, F.: Distribution and innervation of short, interdigitated muscle fibers in parallel-fibered muscles of the cat hindlimb. J. Morphol. **191**(1), 1–15 (1987)
4. Lieber, R.L., Fridén, J.: Functional and clinical significance of skeletal muscle architecture. Muscle Nerve: Offic. J. Amer. Assoc. Electrodiagnostic Med. **23**(11), 1647–1666 (2000)
5. Lichtwark, G., Wilson, A.: Optimal muscle fascicle length and tendon stiffness for maximising gastrocnemius efficiency during human walking and running. J. Theor. Biol. **252**(4), 662–673 (2008)
6. Abe, T., Fukashiro, S., Harada, Y., Kawamoto, K.: Relationship between sprint performance and muscle fascicle length in female sprinters. J. Physiol. Anthropol. Appl. Hum. Sci. **20**(2), 141–147 (2001)
7. Mitsiopoulos, N., Baumgartner, R., Heymsfield, S., Lyons, W., Gallagher, D., Ross, R.: Cadaver validation of skeletal muscle measurement by magnetic resonance imaging and computerized tomography. J. Appl. Physiol. **85**(1), 115–122 (1998)
8. Oberhofer, K., Stott, N., Mithraratne, K., Anderson, I.: Subject-specific modelling of lower limb muscles in children with cerebral palsy. Clin. Biomech. **25**(1), 88–94 (2010)

9. Campbell, R., Wood, J.: Ultrasound of muscle. Imaging **14**(3), 229–240 (2002)
10. Maganaris, C.N.: Force-length characteristics of the in vivo human gastrocnemius muscle. Clinical Anatomy: The Official J. Amer. Assoc. Clinical Anatomists British Assoc. Clinical Anatomists **16**(3), 215–223 (2003)
11. Whittaker, J.L. et al.: Rehabilitative ultrasound imaging: understanding the technology and its applications. J. Orthopaedic Sports Phys. Therapy **37**(8), 434–449 (2007)
12. Ohata, K., Tsuboyama, T., Haruta, T., Ichihashi, N., Kato, T., Nakamura, T.: Relation between muscle thickness, spasticity, and activity limitations in children and adolescents with cerebral palsy. Dev. Med. Child Neurol. **50**(2), 152–156 (2008)
13. Blazevich, A.J., Gill, N.D., Zhou, S.: Intra-and intermuscular variation in human quadriceps femoris architecture assessed in vivo. J. Anat. **209**(3), 289–310 (2006)
14. Finni, T., Ikegawa, S., Lepola, V., Komi, P.: Comparison of force–velocity relationships of vastus lateralis muscle in isokinetic and in stretch-shortening cycle exercises. Acta Physiol. Scand. **177**(4), 483–491 (2003)
15. Noorkoiv, M., Stavnsbo, A., Aagaard, P., Blazevich, A.J.: In vivo assessment of muscle fascicle length by extended field-of-view ultrasonography. J. Appl. Physiol. **109**(6), 1974–1979 (2010)
16. Fornage, B.D., Atkinson, E.N., Nock, L.F., Jones, P.H.: US with extended field of view: phantom-tested accuracy of distance measurements. Radiology **214**(2), 579–584 (2000)
17. Ying, M., Sin, M.-H.: Comparison of extended field of view and dual image ultrasound techniques: accuracy and reliability of distance measurements in phantom study. Ultrasound Med. Biol. **31**(1), 79–83 (2005)
18. Hedrick, W.: Extended field of view real-time ultrasound: a review. J. Diagnos. Med. Sonography **16**(3), 103–107 (2000)
19. Tan, C., Liu, D.C.: Image registration based wide-field-of-view method in ultrasound imaging. In: 2008 2nd International Conference on Bioinformatics and Biomedical Engineering, pp. 2405–2408. IEEE (2008)
20. Bénard, M.R., Becher, J.G., Harlaar, J., Huijing, P.A., Jaspers, R.T.: Anatomical information is needed in ultrasound imaging of muscle to avoid potentially substantial errors in measurement of muscle geometry. Muscle Nerve: Offic. J. Amer. Assoc. Electrodiagnostic Med. **39**(5), 652–665 (2009)
21. Klimstra, M., Dowling, J., Durkin, J.L., MacDonald, M.: The effect of ultrasound probe orientation on muscle architecture measurement. J. Electromyogr. Kinesiol. **17**(4), 504–514 (2007)
22. Kawakami, Y., Ichinose, Y., Fukunaga, T.: Architectural and functional features of human triceps surae muscles during contraction. J. Appl. Physiol. **85**(2), 398–404 (1998)
23. Herbert, R., Gandevia, S.: Changes in pennation with joint angle and muscle torque: in vivo measurements in human brachialis muscle. J. Physiol. **484**(2), 523–532 (1995)
24. Cronin, N.J., Carty, C.P., Barrett, R.S., Lichtwark, G.: Automatic tracking of medial gastrocnemius fascicle length during human locomotion. J. Appl. Physiol. **111**(5), 1491–1496 (2011)
25. Lucas, B.D., Kanade, T.: An iterative image registration technique with an application to stereo vision. Vancouver, British Columbia (1981)
26. Baker, S., Matthews, I.: Lucas-kanade 20 years on: a unifying framework. Int. J. Comput. Vision **56**(3), 221–255 (2004)
27. Gillett, J.G., Barrett, R.S., Lichtwark, G.A.: Reliability and accuracy of an automated tracking algorithm to measure controlled passive and active muscle fascicle length changes from ultrasound. Comput. Methods Biomech. Biomed. Eng. **16**(6), 678–687 (2013)
28. Maïsetti, O., Hug, F., Bouillard, K., Nordez, A.: Characterization of passive elastic properties of the human medial gastrocnemius muscle belly using supersonic shear imaging. J. Biomech. **45**(6), 978–984 (2012)
29. Bouillard, K., Hug, F., Guével, A., Nordez, A.: Shear elastic modulus can be used to estimate an index of individual muscle force during a submaximal isometric fatiguing contraction. J. Appl. Physiol. **113**(9), 1353–1361 (2012)

30. Bouillard, K., Nordez, A., Hug, F.: Estimation of individual muscle force using elastography. PloS One **6**(12), e29261 (2011)
31. Bercoff, J., Tanter, M., Fink, M.: Supersonic shear imaging: a new technique for soft tissue elasticity mapping. IEEE Trans. Ultrason. Ferroelectr. Freq. Control **51**(4), 396–409 (2004)
32. Ford, L., Huxley, A., Simmons, R.: The relation between stiffness and filament overlap in stimulated frog muscle fibres. J. Physiol. **311**(1), 219–249 (1981)
33. Sasaki, K., Toyama, S., Ishii, N.: Length-force characteristics of in vivo human muscle reflected by supersonic shear imaging. J. Appl. Physiol. **117**(2), 153–162 (2014)

Chapter 5
Measurement of Shkeletal Muscle Cross-Sectional Area

Abstract Muscle cross-sectional area (CSA) is related to muscle volume and can be used to study muscle atrophy, while its shape can also change dramatically during muscle contraction. This chapter focuses on several recent reports on the calculation of muscle cross-sectional area under ultrasound, including one report using deep learning.

5.1 Manual Measurement of Cross-Sectional Area (CSA) and Its Relationship with Muscle Strength

The CSA of skeletal muscle is one of the morphological parameters for quantifying muscle function and has been used for the diagnosis of various muscle diseases and muscle assessment in sports medicine [1]. Since representing the maximal number of acto-myosin crossbridges that can be activated in parallel during contraction, the total physiological CSA is directly correlated with the force-generating ability of a given muscle. Therefore, more and more studies have paid attention to the relationship between maximal force and CSA and the estimation of the muscle strength per unit cross-sectional area. It has been understood that the strength must be proportional to the physiological CSA of the muscle [2]. Some differences, however, have been found among the results of research as follows: 4 kg/cm² [2], 6–10 kg/cm ² [3], 6.24 kg/cm² [4, 5], and 9.2 kg/cm² [6].

Ikai et al. conducted a study to determine the strength per unit CSA by means of an ultrasonic method in living human subjects [5]. In addition, they also intended to make clear whether some differences existed among ages and genders as well as trained and untrained subjects. By means of the ultrasonic imaging of the cross-section of the acting muscle bundle, together with the measurement of the muscle strength developed by the subject with maximum effort, the strength per unit area of the muscle was calculated in 245 healthy human subjects, including 119 males and 126 females. The measurement process is described in detail below.

Measurement of Maximum Strength Muscle strength was measured at the arm flexor at the right angle of the elbow joint in the sitting position isometrically. The subject contracted the muscle against a cloth belt attached to the wrist with the

© Springer Nature Singapore Pte Ltd. 2021
Y. Zhou and Y.-P. Zheng, *Sonomyography*, Series in BioEngineering,
https://doi.org/10.1007/978-981-16-7140-1_5

maximum effort. The belt, 45 mm wide, was connected to a strain gauge tensiometer. The legs were extended on a chair. After three trials for measurement, the highest value was adopted as the maximum strength of each individual.

Measurement of the CSA of Muscle The subject was asked to assume a lying position, while his arm was fixed to extend to the bottom of a water tank (Fig. 5.1). An ultrasonic scanner circulated around the upper arm to be tested in 30 s. A pulsed echo was displayed on the cathode ray screen in brightness modulation. The frequency of ultrasonic wave was chosen to be 2.25–5 MHz for a clear view of bone, muscle as well as subcutaneous fat. Figure 5.2 shows the obtained cross-sectional image of the upper arm. In this image, the boundaries among subcutaneous fat, muscle, fascia, and bone can be observed clearly. Under consideration of the structure of subcutaneous fat and fascia, muscle of the upper arm was categorized as flexor and extensor. To measure the size of the tissues, a calibration curve was made by means of Bakelite models of several diameters. The result is summarized as follows:

(1)   The ultrasonic method used in this work was found to be the best way to calculate the CSA of the muscle.
(2)   The arm strength was fairly proportional to the CSA of the flexor of the upper arm regardless of age and gender.
(3)   The strength per unit CSA of flexor muscles of the upper arm was 6.3 kg/cm^2 on average, with standard deviation of 0.81 kg/cm^2. When the CSA of muscle

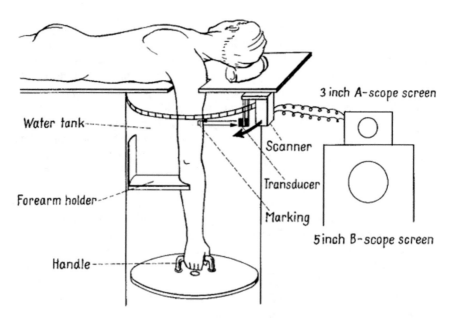

Fig. 5.1 Scheme of the ultrasonic equipment for scanning the upper limb. Reprinted by permission from Springer Nature, Internationale Zeitschrift für Angewandte Physiologie Einschliesslich Arbeitsphysiologie, Calculation of muscle strength per unit cross-sectional area of human muscle by means of ultrasonic measurement, M. Ikai and T. Fukunaga, 1968

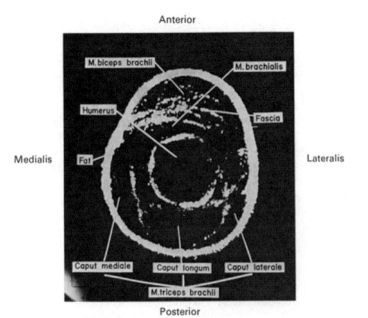

Fig. 5.2 Cross-sectional view of human upper arm by the ultrasonic method. Reprinted by permission from Springer Nature, Internationale Zeitschrift für Angewandte Physiologie Einschliesslich Arbeitsphysiologie, Calculation of muscle strength per unit cross-sectional area of human muscle by means of ultrasonic measurement, M. Ikai and T. Fukunaga, 1968

was measured at the extensive position of the forearm, the strength per unit area was calculated to be 4.7 kg/cm^2 at the flexed position of the forearm.

(4) As to the individual variation, the strength per unit area was distributed over a range of from 4 to 8 kg/cm^2.

(5) The strength per unit CSA was almost the same in male and female subjects regardless of age. In addition, no significant difference was found in ordinary and trained adults.

Manual measurement has been the conventional method for obtaining CSA from ultrasound images. However, it is time-consuming and labor intensive and its result depends very much on the experience of operators in segmentation. Moreover, this static measurement method still cannot clearly explore the mechanism behind the continuous changes of CSA during dynamic contraction. Recently, computer-assisted design software has been applied for medical image segmentation. Chen et al. [7] proposed an image-tracking algorithm, namely constrained mutual-information-based free-form deformation (C-MI-FFD) tracking, to automatically extract the CSA of the rectus femoris (RF) muscle from continuously captured US images. In the proposed C-MI-FFD method, they aimed to determine the transformation function which described the deformation between two successive images by minimizing a MI-based objective function. To further improve the tracking performance, they incorporated three image structure-derived constraints, including smoothing constraint, feature point constraint, and edge constraint, into the MI objective function. The MI-based free-form deformation method was originally proposed in [8] to do shape registration in 2D space based on the mutual information between two images and it was extended in this realization for object tracking. More precisely, the intensities of two images, say A and B, to be matched were treated as random variables with probability density functions (pdfs), $p_A(i_A)$ and $p_B(i_B)$, and joint pdf $p_{AB}(i_A, i_B)$. B was then deformed by means of a transformation function $T(B)$ with parameters to be determined. Ideally, when the two images are registered, the mutual information between A and $T(B)$ is maximized, as shown in the following equation,

$$MI = I(A, T(B)) = H(p_A(i_A)) + H\big(p_{T(B)}(i_B)\big) - H\big(P_{A,T(B)}(i_A, i_B)\big)$$

$$= \iint\limits_{I(A), I(B)} p_{A,T(B)}(i_A, i_B) \log \frac{p_{A,T(B)}(i_A, i_B)}{p_A(i_A) p_{T(B)}(i_B)} di_A di_B \qquad (5.1)$$

where i_A and i_B are the intensity valuables of A and $T(B)$, $H(p_A(i_A))$, $H(p_{T(B)}(i_B))$, $H(p_{A,T(B)}(i_A, i_B))$ are, respectively, the entropies of A, $T(B)$ and their joint entropy. Therefore, by maximizing $I(A, T(B))$ using the parameters of $T(\cdot)$, the two original images can be registered and the segmented object can be tracked continuously. In practice, the pdfs are approximated by the kernel method, which can be computed directly from the image data. However, for the highly noisy ultrasound images under study, the conventional MI-FFD method cannot achieve satisfactory tracking performance. For more accurate matching, the transformation in the proposed C-MI-FFD method was carried out in two steps: (1) global transformation and (2) local transformation. In the first step of global transformation, the parameters were determined by matching the two images globally so as to model their relative scales, translations, and rotations. The global deformation function T_G is chosen as an affine transformation $T_G = \psi[x, y, 1]^T$, where x and y are the coordinates of the images and ψ is a 3×3 affine transformation matrix in the homogeneous coordinate that can be obtained by maximizing Eq. 5.1. Let B' be the transformed image obtained by the affine transformation after the global transformation. Further, B' will be refined by the subsequent local transformation. In the second step of local transformation, the local deformation $T_L(B')$ is defined by a 2D spline function.

The transformation parameters, which are the displacement values at a regular grid to interpolate the spline function, are, again, determined by maximizing the MI objective function in Eq. 5.1. The local transformation TL(B') is parameterized by the displacement vectors at a uniform grid of control points,

$$C, p_c(m, n) = \big[p_{c,\bar{x}}(m, n), p_{c,\bar{y}}(m, n)\big]^T$$

which are indexed by m = 1, …, M, n = 1, …, N. If (X, Y) is the resolution of the input image, the spacing of the control points in the $\bar{x}$ and $\bar{y}$ directions are

$$\Delta \bar{x} = X/M$$

and

$$\Delta \bar{y} = Y/N$$

respectively. The deformation of any pixel in the image is obtained by spline interpolation of those at the grid points C. Therefore, the deformation of pixel

$$(i, j),\ P(i, j) = \left[p_{\bar{x}}(i, j),\ p_{\bar{y}}(i, j) \right]^{T}$$

where $1 \leq i \leq X,\ 1 \leq j \leq Y$ can be written as Eq. 5.2:

$$P(i, j) = \sum_{\mu=0}^{3} \sum_{\gamma=0}^{3} \beta_{\mu}(u)\beta_{\gamma}(v) p_{c}(m + \mu, n + \gamma) \tag{5.2}$$

where $\beta_{\mu}(u)$ and $\beta_{\gamma}(v)$ are, respectively, the cubic B-spline function with,

$$u = \frac{i}{\Delta_x} - \left| \frac{i}{\Delta_x} \right|$$

$$v = \frac{j}{\Delta_y} - \left| \frac{j}{\Delta_y} \right|$$

and

$$\{(m + \mu, n + \gamma) | (\mu, \gamma) \in [0, 3]\}$$

are the neighboring control points of (i, j). Then, the local transformation of B' can be obtained using Eq. 5.3:

$$T_L\big(B'(i, j)\big) = B'\big(i + P_{\bar{x}}(i, j),\ j + P_{\bar{y}}(i, j)\big) \tag{5.3}$$

By substituting (5.2) into (5.1), one gets the local matching objective function $E_{local} = -MI$ to be minimized.

The global transformation only needs to estimate nine model parameters in the 3×3 matrix ψ, but the local transformation needs to estimate a large number of parameters (i.e., each pixel has two parameters to be estimated in two directions). In order to reduce the variance of the local transformation parameters, some prior constraints, including the smoothing constraint, the feature point constraint, and the edge constraint, are added into the local matching objective function. In the proposed

C-MI-FFD method, Chen et al. [7] incorporated several constraints, which were derived from structural information of the images, into the MI objective function so as to improve the tracking performance. The details of these constraints are elaborated as follows.

(1)   **Smoothing constraint**. A popular smoothing constraint is the L_2 norm of the displacement as shown in Eq. 5.4:

$$E_{smooth} = \sum_{(m,n)} \left(\frac{\partial P_c(m, n)}{\partial m}^2 + \frac{\partial P_c(m, n)}{\partial n}^2 \right) \tag{5.4}$$

(2)   **Feature point constraint**. If the pair of images being registered does have distinct geometrical features as correspondences, incorporating this geometrical feature information can greatly improve accuracy and efficiency. In this algorithm, the scale invariant features [9] can be used as a feature constraint. When feature points and correspondences are available, these constraints can be conveniently integrated into the registration framework. Assuming that the total number of features is K, and for each feature, there is a pair of corresponding points, $I_k^{(T)}, k = 1, \ldots, K$, on the target image and $I_k^{(S')} = T_L\left(I_k^{(S)}\right), k = 1, \ldots, K$, on the locally transformed source image, then the following term can be incorporated as a feature constraint (Eq. 5.5):

$$E_{feature} = \sum_{k=1}^{K} D\left(I_k^{(S')} - I_k^{(T)}\right) \tag{5.5}$$

where $D\left(I_k^{(S')} - I_k^{(T)}\right)$ is the Euclidean distance between $I_k^{(T)}$ and $I_k^{(S')}$.

(3)   **Edge constraint**. The edge term is expressed as Eq. 5.6:

$$E_{edge} = \sum_{i=1}^{M} \sum_{j=1}^{N} \left[A_{edge}(i, j) - B_{edge}^{(T)}(i, j) \right] \tag{5.6}$$

where A_{edge} and $B_{edge}^{(T)}$ are the edge maps of A and $T(B)$, respectively, and M and N are the height and width of the images. Color tensor-based edge detection was used in this study to extract the image edges A_{edge} and $B_{edge}^{(T)}$.

With all these constraint terms, the final local objective function to be minimized is expressed as Eq. 5.7:

$$E_{local} = -MI + E_{smooth} + E_{feature} + E_{edge} \tag{5.7}$$

To solve the nonlinear optimization problem, an L-BFGS algorithm [10] was used in the proposed C-MI-FFD method. An advantage of the L-BFGS algorithm is that explicit evaluation of the Hessian matrix is not required, so it can be recursively estimated. Moreover, the L-BFGS algorithm is much faster than the conventional level set method in [8] for solving the nonlinear optimization problem. For each measurement, the first image in the sequence was selected as reference and the boundary of the RF muscle was outlined with smooth lines by the investigator using ImageJ software (ImageJ, National Institutes of Health, USA). Then the C-MI-FFD method was applied to track the CSA boundaries in the subsequent images. The CSA value was then normalized as the percentage decrease DCSA, with respect to the value at the relaxed condition. To evaluate the reliability of the automatic image tracking algorithm, a total of 100 randomly selected ultrasound images were measured twice by the image processing algorithm and the investigator. The intra-class correlation coefficient (ICC) based on a one-way random model and standard error of the measurement (SEM) were calculated [11].

Four images of a typical trial, including the first image, the first image with a manually drawn boundary, the image at 50% MVC, and the image at 50% MVC with automatically tracked boundary, are shown in Fig. 5.3. The ICC for CSA measurements of the RF muscle in this study was 0.987 ($P < 0.0001$), and the SEM was 0.15 cm^2. The average CSA for the nine subjects was 6.15 ± 0.48 cm^2 (mean $\pm$ SEM) under the relaxed condition.

5.2 Deep Learning

Deep learning is a kind of machine-learning method based on a deep neural network. It has been proved to be an effective method in medical image analyses such as image registration, segmentation, and computer-aided disease diagnosis or prognosis. For the measurement task of CSA, Chen et al. [12] developed an automatic segmentation method for RF muscles in US images based on a CNN. This end-to-end method was used to track the CSA change during muscle contraction. Figure 5.4 shows the architecture of the CNN used in this study. This network consisted of two phases, namely feature representation and score map reconstruction phases.

Feature Representation Phase The feature representation phase consisted of five down-sampling blocks. A down-sampling block included a convolution layer, an activation layer, and a pooling layer. The convolution layer detected local features from the input images, and it was configured to ensure that the size of the output feature map was equal to that of the input. The kernel size, stride, and pad were 3, 1, and 1, respectively. After each convolution layer, Chen et al. [12] used the rectified linear unit function as the activation function of the activation layer. In all five down-sampling blocks, the convolution layers were followed by a pooling layer. They applied pooling with 2×2 filters and two strides, which decreased the dimensionality of the feature map by 50%. The input images with a size of 640×480 pixels were transformed into 20×15-pixel feature maps in the feature representation phase.

(a)   **(b)**

(c)   **(d)**

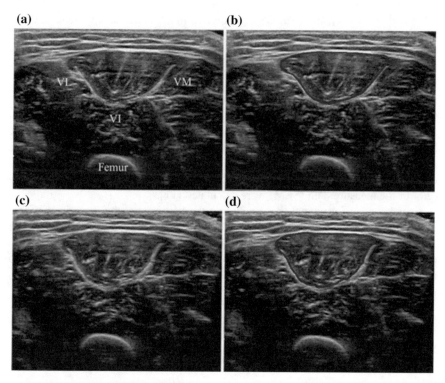

Fig. 5.3 Ultrasound images of the RF muscle in one trial [7]. **a** The first image in the image sequence, where the neighboring quadriceps muscles [vastus lateralis (VL), vastus intermedius (VI), and vastus medialis (VM)] are labeled. **b** The first image in the image sequence, with a manually outlined boundary as a reference for further image processing. **c** The image at 50% MVC. **d** The image at 50% MVC, with an automatically outlined boundary by the proposed C-MI-FFD algorithm. Reprinted by permission from Springer Nature, European journal of applied physiology, Sonomyographic responses during voluntary isometric ramp contraction of the human rectus femoris muscle, X. Chen, Y.-P. Zheng, J.-Y. Guo, Z. Zhu, S.-C. Chan, and Z. Zhang, 2012

Score Map Reconstruction Phase The main purpose of the score map reconstruction phase was to reconstruct the score map into the same size of input images by up-sampling. This reconstruction phase consisted of five up-sampling blocks, each of which was composed of a deconvolution layer, a concatenation layer, a convolution layer, and an activation layer. The deconvolution layer was responsible for enlarging images during the up-sampling process. In this phase, the size of the feature map was enlarged from 20×15 to 640×480 pixels by five deconvolution layers. The concatenation layer connected the enlarged score map with the corresponding feature map and fused the feature maps in the previous pooling layers or convolution layers with the current feature maps in the deconvolution layer. In this operation, the ratio of the number of the enlarged score map to the feature map was set as 1:1. This skip-layer design could capture more multiscale contextual information to improve the accuracy of segmentation. The convolution layer integrated the image information

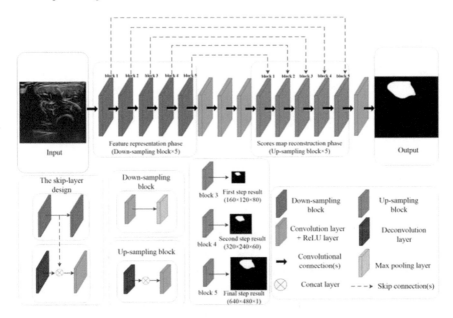

Fig. 5.4 Block diagram showing the architecture of the RF muscle segmentation CNN. Reprinted from J. P. Weir, "Quantifying test–retest reliability using the intraclass correlation coefficient and the SEM," The Journal of Strength and Conditioning Research, vol. 19, no. 1, pp. 231–240, 2005, with permission from WILEY

after the concatenation layer. To fuse the low- and high-resolution information pixel by pixel, the kernel size of the last convolution layer was set as 1×1. With all of the up-sampling blocks, the score maps were reconstructed to an output image with the same pixel size of 640×480 as the input image. To optimize the network, Chen et al. estimated the loss by calculating the Euclidean distance between the ground truth and the reconstructed score map [11]. Then, the network parameters were iterated and upgraded by backpropagation from the loss.

In this study, three male and two female volunteers participated in the data collection, with a mean age ($\pm$SD) of 30.5 ± 1.5 years, body weight of 62.5 ± 3.3 kg, and height of 165.5 ± 2.5 cm. Table 5.1 shows the performance metrics used for evaluating the proposed method. The mean precision, recall, and DSC values were 0.936 ± 0.029, 0.882 ± 0.045, and 0.907 ± 0.023, respectively, showing high accuracy in obtaining the CSA of the RF muscle from US images. The performance of the proposed method was also compared with a state-of-the-art CSA tracking algorithm: C-MI-FFD tracking. The same testing dataset was used to evaluate the segmentation performance of the C-MI-FFD method, and the results are also shown in Table 5.1. Repetitive experiments showed that the proposed method performed almost as well as when undertaking the same task manually. The performance of the proposed CNN method with low standard deviations across the repetitions demonstrated that the method has great potential for clinical applications. Meanwhile, the proposed method was shown to be better than the C-MI-FFD method in terms of the DSC, precision,

Table 5.1 Experimental results of the deep learning method and the performance comparison with C-MI-FFD method

Experiment	Precision	Recall	DSC
1	0.951	0.914	0.932
2	0.977	0.811	0.885
3	0.908	0.864	0.885
4	0.940	0.917	0.929
5	0.906	0.905	0.906
Mean	0.936	0.882	0.907
SD	0.029	0.045	0.023
C-MI-FFD	0.929	0.725	0.814

and recall. Figure 5.5 shows a comparison between the RF muscle contours extracted by the proposed deep learning method and the C-MI-FFD method. The C-MI-FFD method can be used to track muscle morphologic changes by shape transformation in a 2-dimensional space based on mutual information between two images. Therefore, the C-MI-FFD method is a semiautomatic method. The boundary of the RF muscle needs to be manually segmented as a reference in the first image frame before the tracking procedure. Compared with the C-MI-FFD method, the accuracy of this deep learning method greatly improved the working efficiency of the segmentation procedure in obtaining the CSA of the RF muscle, and this proposed end-to-end method achieved enhanced performance in the speed of processing images (Table 5.1).

5.3 Summary

The results of previous studies have demonstrated that skeletal muscle cross-sectional area is closely correlated with muscle strength; thus, CSA evaluation has great value for the functional assessment of muscles. Various methods including deep learning approaches have been adopted to detect CSA in muscle ultrasound images. With the development of more reliable, accurate and efficient detection methods, the cross-sectional area of muscle can be obtained in real-time during contraction. This will allow researchers studying the correlation between the muscle strengths with change of muscle CSA during contraction. With the development of wearable ultrasound imaging systems, dynamic CSA measurement can soon be achieved in daily activities for different subjects, such as during walking and running. For all these applications, the approaches for attaching ultrasound probes onto the body should also be well addressed, as the CSA can be easily changed due to the compression applied by the attached probe. Methods for proper attachment of ultrasound probes as well as the development of ultrasound probes that are suitable for attachment is another important research topic in muscle ultrasound imaging in the future. It should be noted that most of the abovementioned methods can also output muscle contour (not

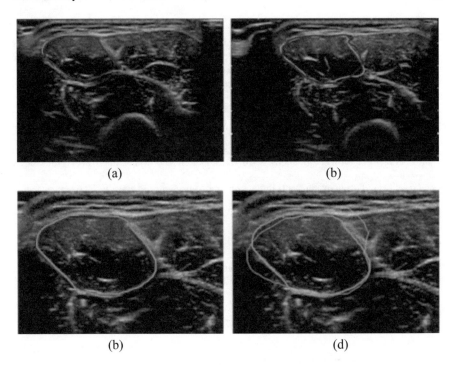

(a)   (b)

(b)   (d)

Fig. 5.5 Two typical tracking examples [12]. **a** The result obtained by the deep learning method for an ultrasound image of one muscle, and **b** the result obtained in a consequent image using the same method; **c** the result of deep learning method for an image of another muscle, and **d** the result obtained by the constrained MI-FFD tracking method for the image same as (c). The ground truth obtained by manual segmentation, tracking result obtained using the deep learning method, and tracking result obtained by the C-MI-FFD method are represented by green, red, and yellow lines, respectively. Reprinted from X. Chen, C. Xie, Z. Chen, and Q. Li, "Automatic Tracking of Muscle Cross-Sectional Area Using Convolutional Neural Networks with Ultrasound," Journal of Ultrasound in Medicine, vol. 38, no. 11, pp. 2901–2908, 2019., with permission from WILEY

just its area) in the transverse cross-sectional plane, which can be further explored to relate muscle contraction activity with such architectural changes of muscle contour.

References

1. Faltus, J., Boggess, B., Bruzga, R.: The use of diagnostic musculoskeletal ultrasound to document soft tissue treatment mobilization of a quadriceps femoris muscle tear: a case report. Int. J. Sports Phys. Ther. **7**(3), 342 (2012)
2. Hettinger, T.: Physiology of Strength. Pickle Partners Publishing (2017)
3. Fick, R.: Handbuch der Anatomie und Mechanik der Gelenke unter Berücksichtigung der bewegenden Muskeln. BoD–Books on Demand (2013)
4. Hermann, L.: Zur Messung der Muskelkraft am Menschen. Nach Versuchen von cand. med. C. Hein und von Dr. med. Th. Siebert. Von. Archiv für die gesammte Physiologie des Menschen

und der Thiere **73**, 429 (1898)

5. Ikai, M., Fukunaga, T.: Calculation of muscle strength per unit cross-sectional area of human muscle by means of ultrasonic measurement. Internationale Zeitschrift für Angewandte Physiologie Einschliesslich Arbeitsphysiologie **26**(1), 26–32 (1968)

6. Morris, C.B.: The measurement of the strength of muscle relative to the cross section. Res. Q. Am. Assoc. Health Phys. Educ. Recreation **19**(4), 295–303 (1948)

7. Chen, X., Zheng, Y.-P., Guo, J.-Y., Zhu, Z., Chan, S.-C., Zhang, Z.: Sonomyographic responses during voluntary isometric ramp contraction of the human rectus femoris muscle. Eur. J. Appl. Physiol. **112**(7), 2603–2614 (2012)

8. Huang, X., Paragios, N., Metaxas, D.N.: Shape registration in implicit spaces using information theory and free form deformations. IEEE Trans. Pattern Anal. Mach. Intell. **28**(8), 1303–1318 (2006)

9. Lowe, D.G.: Distinctive image features from scale-invariant keypoints. Int. J. Comput. Vision **60**(2), 91–110 (2004)

10. Byrd, R.H., Lu, P., Nocedal, J., Zhu, C.: A limited memory algorithm for bound constrained optimization. SIAM J. Sci. Comput. **16**(5), 1190–1208 (1995)

11. Weir, J.P.: Quantifying test-retest reliability using the intraclass correlation coefficient and the SEM. J. Strength Conditioning Res. **19**(1), 231–240 (2005)

12. Chen, X., Xie, C., Chen, Z., Li, Q.: Automatic Tracking of Muscle Cross-Sectional Area Using Convolutional Neural Networks with Ultrasound. J. Ultrasound Med. **38**(11), 2901–2908 (2019)

Chapter 6
Enhancement of Ultrasound Images of Muscles

Abstract It is well-known that there can be quite serious noises in ultrasound images and ultrasound imaging is sensitive to the orientation of the probe during imaging. Therefore, there are often problems that the image quality is not satisfactory or the image content in the context frames fluctuates greatly, which leads to difficulties in the dynamic assessment of muscles during their contraction. This chapter introduces some research works to enhance ultrasound muscle images, especially one using a deep learning technique.

Medical images contain extremely rich pathological information and play an important role in clinical diagnosis and treatment by doctors. Since the mid-1980s, medical ultrasound imaging, X-ray computed tomography (CT), magnetic resonance imaging (MRI), and radioisotope imaging (SPECT) have been recognized as the four major imaging technologies in modern medicine. And due to ultrasound imaging having the properties of low cost, noninvasiveness, no radiation, real-time, and rapid diagnosis, it, especially B-mode ultrasound imaging, has been widely used in the prevention, diagnosis, and screening of different human diseases. It has become one of the most widely used and routine diagnostic tools in modern medicine.

When traditional medical ultrasonic diagnostic equipment is being operated, the transducer sends ultrasonic pulses to the human body. The frequency of the medical detection ultrasonic pulse is generally 2.5–13 MHz, but the higher the ultrasound frequency, the weaker the diffraction, and the higher the imaging resolution. At the same time, the higher the frequency, the faster the sound wave is attenuated and the penetration depth becomes smaller. After the pulse enters the acoustic interface of organs, different refraction, diffraction and reflection waves, and refraction and reflection waves will be generated. Ultrasound echo carries part of the energy back to the transducer, and is converted from sonic energy into electrical energy by the transducer, and displays the corresponding organ or tissue shape on the display after being processed, such as envelope detection, etc.

However, the images obtained from the machine are affected by many factors such as imaging equipment and inspection methods, and there are interferences, such as affected resolution, artifacts, noise, etc., resulting in people with different abilities or different backgrounds drawing different conclusions about the same medical image.

Y. Zhou and Y.-P. Zheng, *Sonomyography*, Series in BioEngineering,
https://doi.org/10.1007/978-981-16-7140-1_6

Therefore, it is very necessary to use a computer to enhance image quality, including ultrasound imaging for muscles.

6.1 Imaging Characteristics of Ultrasound Images

The unique advantages of ultrasound imaging include:

(1) It can better display muscle, soft tissue, and bone surface.
(2) It can present real-time images, and the operator can select the most useful part as the basis for diagnosis, usually for quick diagnosis. Real-time ultrasound imaging can also provide guidance for conducting biopsies and injections more easily.
(3) Ultrasound imaging is recognized as a safe imaging method, and the World Health Organization's evaluation is: "Diagnostic ultrasound is considered to be a safe, effective and highly flexible imaging method that can provide a large body in a fast and cost-effective manner. Clinically relevant information on some parts."
(4) Ultrasound equipment is widely available and relatively flexible, providing a small, portable scanner with no location restrictions.
(5) It is cheaper and faster than other imaging modes, such as CT and MRI.

Meanwhile, ultrasound imaging also has some shortcomings, including:

(1) Ultrasound imaging does not perform well when there is air between the probe transducer and the organ. For example, during rapid exercise, there is no guarantee that the probe can be in close contact with the skin, which may have an adverse effect of poor imaging.
(2) There are different artifacts and noises in ultrasound images.
(3) The depth of ultrasonic penetration is limited by the frequency of imaging. High fat content can affect imaging quality and diagnostic accuracy.
(4) The method depends largely on the experience of the operator. A high level of scanning skill and experience is required to obtain high-quality images and make accurate diagnoses.

6.2 Ultrasound Image Noise Analysis

Gaussian noise is generated during operation of the transducer of ultrasonic devices, such as amplifier. Gaussian noise is an additive noise, which is generally formed by the superposition of thermal noise generated by capacitors, resistors, integrated circuits, etc., and electromagnetic interference noise generated by other electrical equipment. It is characterized by a power spectral density that is independent of the pixel points of the ultrasound image and the signal distribution. With the continuous advancement of technology and continuous improvement of equipment, engineers have been able to continuously reduce the impact of Gaussian noise on ultrasound images.

However, in the ultrasonic image acquisition process, we use the reflection, scattering, and refraction characteristics of ultrasonic beams. Due to the non-uniformity of various parts of human tissue and the uncertain nature of spatial distribution, when ultrasonic waves are transmitted into the human body, a large number of randomly distributed scattered particles are formed, and the interaction between the scattered particles produces correlated scattered beams. In the echo reflection process, due to the interference effect of the reflected echoes and the mutual interference between the scattered beams, when the echoes of different beams overlap, the addition and subtraction of the amplitudes may occur due to the different phases of the echoes, leading to speckle noises in ultrasound images.

From the perspective of ultrasound image quality, the main influencing factor is speckle noise. Its existence not only masks the characteristics of low-contrast tissue, but also greatly reduces the ability of observers to analyze detailed features, which seriously affects the quality of ultrasound images. Even experienced operators may have difficulties, and subsequent processing of images entails great challenges. Therefore, in order to improve the quality of images, they must be processed, including those for muscles. However, in the field of medical ultrasound, it is widely believed that the distribution and shape of speckle noise in an ultrasonic image contain information useful for diagnosis, and therefore, in an application for improving image quality, such as image contrast enhancement, texture enhancement, etc., image detail information should be preserved. Under this premise, speckle noise should not be excessively eliminated. Therefore, the characteristics of this special speckle noise are studied, and on this basis, how to suppress speckle noise has become an important topic, and the image edge and detail features are preserved and enhanced for accurate edge detection, image recognition, segmentation and localization, and feature extraction.

6.3 Ultrasound Image Enhancement Methods

In the process of image generation, transmission, and transformation, due to various factors, images often have some differences from the original object or between original images, which causes difficulties and inconveniences for obtaining effective information. Sometimes the acquired image is dark overall and the contrast is low, making it difficult for operators to extract valid information from it. The purpose of image enhancement is to improve the visual effect of the image or enhance some features of the image according to the characteristics of the image, the existing problems or the purpose of the application, so that the image is converted into a form more suitable for human and machine processing and analysis. The image is thus clearer for human eye observation and provides as much information as possible for diagnosis.

Image enhancement currently lacks a unified theory, which is related to the absence of a universal, objective standard for measuring image enhancement quality. Image enhancements cover a wide range of content, including noise reduction, feature enhancement, and image quality improvement. More details are elaborated as follows.

6.4   From the Perspective of Image Denoising

After years of development, a variety of filtering algorithms have been applied to ultrasonic image processing. According to the working principle of these filtering algorithms, they can be roughly divided into spatial domain-based filtering algorithms, transform domain-based filtering algorithms, and diffusion theory filtering algorithms.

The principle of the spatial domain filtering algorithm is to use various image filtering window templates to smooth the image or to adjust the pixel values according to the local statistical characteristics of the image to achieve the purpose of suppressing noise. There are many such filtering algorithms, such as median filtering, Lee filtering, Frost filtering, and non-local means filtering.

The transform domain filtering algorithm can be further divided into two categories: filtering algorithms based on frequency domain transform and filtering algorithms based on wavelet domain transform. The frequency domain transform filtering algorithm is mainly determined by performing Fourier transform on the image of the transform window. The original image is first decomposed into images with different frequency ranges of noise and uncontaminated images, then the appropriate frequency domain bandpass filter is applied for filtering to remove the frequency domain noise, and finally inverse transformation of the Fourier transform is used to obtain the denoised image. Its typical representative is the homomorphic filtering algorithm. The wavelet transform filtering algorithm uses wavelet filtering theory to divide an image into a series of images representing different scale information and wavelet information, denoising the low-resolution image containing low-frequency components, and retaining the main information of the image, including high-frequency components. The high-resolution image is processed with appropriate domain values to preserve the main edge information and discard the high-frequency noise, and finally the processed image is reconstructed. This filtering algorithm is more complicated to implement, but the performance of noise suppression and detail retention is superior.

The filtering algorithm based on diffusion theory includes traditional P-M anisotropic diffusion and corresponding improved algorithm. This type of algorithm can effectively suppress noises while preserving image details, and even enhance contrast and improve image quality. The basic idea is to generate the corresponding spatial scale images by solving the partial differential diffusion equations whose initial values are the original images, adopting different diffusion coefficients and iteration times, and then generating the final filtered images by each scale image transformation iteration. The result is not affected by the filtering window and shape, but it has the problem of choosing the appropriate diffusion coefficient and time step for each region of the image.

6.5 From the Perspective of Improving Image Contrast

Histogram equalization is a classic and effective image enhancement method. Although the method is simple and computationally fast, it does the same for all pixels and does not take into account the local features of an image. Adaptive histogram equalization also has uneven contrast enhancement, the enhanced image looks unnatural, and the noise is enhanced to the same extent. According to the characteristics of medical images, local enhancement is undoubtedly a good method. One example of local enhancement is to use the square root function in the processing algorithm to make dark image regions brighter, but this approach also enhances noises and the background in the image. Therefore, applying the traditional local contrast enhancement method to ultrasound images does not yield satisfactory results. The classical Laplacian method has the characteristics of protruding edges, but it is more sensitive to isolated points. The image enhanced by the Laplacian method is very grainy and enhances noise.

There are also other common enhancement methods, such as the Gabor transform and the top-hat transform, which increase the contrast between the ROI and the background, but they may also simultaneously enhance the speckle noise patterns and even add new artifacts. Figure 6.1 shows a typical example of an ultrasound image before and after the processing. It can be observed that the muscle fiber structures in the image are greatly enhanced; however, some speckle noise patterns are also enhanced; thus, some artifacts are added to the processed image.

Overall, in the enhancement of ultrasound images of muscle, it is still a challenging task to balance the change of image structural information, remove speckle noise, and improve the contrast between ROI and background. More research works are needed in this area.

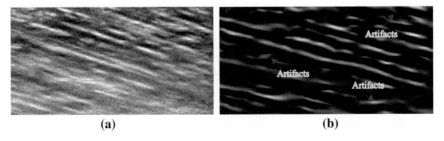

(a)   **(b)**

Fig. 6.1 Ultrasound image of muscle before and after image enhancement. **a** The original image; and **b** enhanced image

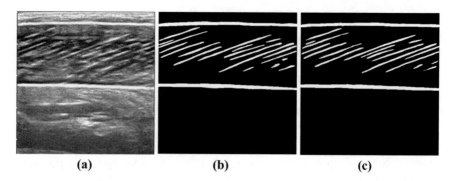

Fig. 6.2 A typical ultrasound image of muscle and the results after processing. **a** The original image, **b** the image manually labeled for muscle fibers, and (c) segmentation result using the deep learning method

With the continuous development of image processing techniques, very good results have been achieved of using deep learning for the segmentation and feature detection of medical images. One example of such application is described here. It has been noted that human eyes can distinguish whether intermittent muscle fibers belong to the same muscle fiber in an ultrasound image or not. Inspired by this observation, experts working in the field related to muscle assessment were approached to manually label the fiber structures in gastrocnemius muscle images (Fig. 6.2a, b). Then the result was used as a training reference for the deep learning algorithm (Fig. 6.2c).

A ratio of data of 7:3 was used to allocate images into the training set (735 images) and test set (315 images), and the training set was amplified, including rotation, change of image gray level, increase or decrease of Gaussian noise, etc., and then the U-net convolutional neural network was used to segment the muscle ultrasound images. During the experiment, the data were not pre-processed by any processing method. Such a method can complement the intermittent muscle fibers in the ultrasound image and segment the fascia and muscle fibers. Good performance was achieved in comparison with the manual results obtained by the experts, with dice value of 89.1%, accuracy of 96.9%, precision of 85.2%, sensitivity of 85.8%, specificity of 98.24%, and AUC of 92.02%. As we can observe from Fig. 6.2, almost all the muscle fiber features identified by manual detection were obtained by the deep learning method. However, a larger dataset is required to fully demonstrate the potential of this new method. Further research works are also needed to understand how the richness of characteristics of the images affects the generalization of this method.

6.6 Summary

In summary, image enhancement and feature extraction for ultrasound images of muscles are still challenging topics. One important feature is the muscle fiber structure, and this should be the focus for future research in this area. If the fiber features can be extracted automatically from ultrasound images in real time, muscle functions can be analyzed dynamically. It has been demonstrated, though at its early stage, that deep learning for muscle fiber extraction has great potentials in this area. In comparison with traditional image enhancement (such as MVEF or Gabor filtering, as introduced in Sect. 2.4.2) for suppressing speckle noises in ultrasound images of muscle, the deep learning method to extract muscle fiber features shows better performance in regard to the target features, i.e. muscle fiber structure. Considering that the required features and parameters in muscle images are well defined, such as thickness, cross-section area, pennation angle, fiber orientation and length, etc., feature-based image processing will be particularly useful for muscle ultrasound images. In future studies, researchers can target to develop automatic methods to extract all the required features of muscle for muscle ultrasound images using advanced approaches, such as deep learning.

Chapter 7
Ultrasound Image Analysis Using AI

Abstract In recent years, with the rapid development of deep learning methods, neural networks have been proven to have great potentials in the field of medical image analysis. Certainly, such technical advancements in artificial intelligence also enable some new possibilities in the investigation of skeletal muscle using ultrasound imaging, both for existing or new topics. In this chapter, two examples related to this advancement are introduced.

7.1 Analysis of Muscle Fiber Orientation with Deep Learning Methods

In 2018, Cunningham et al. presented an investigation into the feasibility of using deep learning methods for developing arbitrary full spatial resolution regression analysis of B-mode ultrasound images of human skeletal muscle, and to be specific, of muscle fiber orientation [1]. In Chap. 4, methods regarding pennation angle and fiber orientations are regarded as inadequate, as they often require manual region selection and feature engineering, providing low-resolution estimations (one angle per muscle) and deep muscles are often not attempted. Cunningham and co-workers used deconvolutions and Max-Unpooling (DCNN) to regularize and improve predicted fiber orientation maps for the entire image, including deep muscles, removing the need for automatic segmentation and compared their results with those with a plain convolutional neural network (CNN) and deep residual convolutional network (ResNet), as well as a previously established feature engineering method, for the same task. Dynamic ultrasound image sequences of the calf muscles were acquired (25 Hz) from eight healthy volunteers (four males, ages: 25–36, median 30). A combination of expert annotation and interpolation/extrapolation provided labels of regional fiber orientation for each image. Neural networks (CNN, ResNet, DCNN) were then trained both with and without dropout using the approach of leave one out cross-validation. The results demonstrated robust estimation of full spatial fiber orientation within approximately a 6° error, which was an improvement over some previous methods. Figure 7.1 shows the results obtained by different neural network methods.

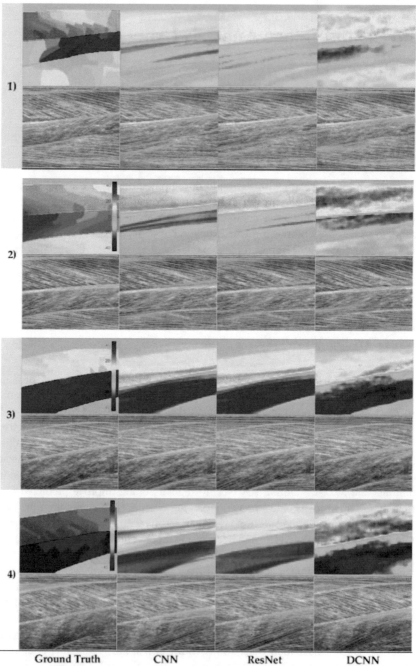

◄**Fig. 7.1** Representative neural network predictions. Rows 1–4 show four respective test participants in the dataset. In each row, the top image shows a fiber orientation heat map and the bottom image shows a line trace representation of the heatmap, overlaid on the ultrasound image. Columns show respectively, ground truth, as well as results obtained by different deep learning methods. Reprinted with permission from R. Cunningham, M. B. Sánchez, G. May, and I. Loram, "Estimating Full Regional Skeletal Muscle Fibre Orientation from B-Mode Ultrasound Images Using Convolutional, Residual, and Deconvolutional Neural Networks," Journal of Imaging, vol. 4, no. 2, p. 29, 2018. https://doi.org/10.3390/jimaging4020029

It was concluded that the application of DCNNs to this problem has opened new potential for high-resolution analysis of skeletal muscle, from prediction of strain and motion maps to segmentation of muscles and other structures of interest. This report provides further evidence that deep learning methods can surpass state-of-the-art performance, even when there is not an abundance of labeled data available, just by organizing the architecture (deconvolutions) of the network.

7.2 Automatic Classification of Male and Female Skeletal Muscle Using Ultrasound Imaging

Besides segmentation tasks, deep learning has also propelled the classification task previously in medical image analysis. An important and very tempting topic in skeletal ultrasound is the automatic classification of muscle status, such as muscle condition for training of athletes, progress assessment of muscles in rehabilitation, diagnosis of muscle health, and so on. Though this topic is still to be explored and expecting progress, Zhou et al. reported a different classification task in 2019 [2] for the automatic classification of skeletal muscle from different genders using ultrasound imaging.

Differences exist between skeletal muscle of females and males, including strength, contractile speed, energy metabolism, fiber type composition, etc. [3]. These differences may lead to at least one phenomenon that women tend to fall two times more than men in aged populations [4]. It is important to study and visualize the skeletal muscle differences between males and females to provide precision medicine and improved health planning [5, 6]. In the past decade, there have been different techniques developed to image the muscle structure parameters such as pennation angle, fascicle length, muscle thickness, and CSA [5, 7–11]. Knowledge regarding the skeletal muscle differences between men and women has been extended in recent studies. For example, differences were explored in terms of stiffness [12]. In addition, the possible differences between B-mode ultrasound images of skeletal muscle from male and female subjects have been investigated [10]. Based on comparisons of muscle measures for both the spatial and frequency domains, the differences were reported in several scenarios of muscle contraction and posture combinations [10].

However, it remains unknown if the differences are significant enough for physicians to visually distinguish them, by a computer directly from ultrasound images. Furthermore, previous methods for qualitative muscle image analysis were heavily based on manual reading or processing [13–15]. With respect to automating this analysis, popular deep learning techniques in the artificial intelligence community [16, 17] are achieving exciting performance results in pattern classification tasks [18]. Unfortunately, there is no report yet on the classification of ultrasound images of skeletal muscle from different genders using deep learning. Therefore, in the report by Zhou's team, two questions were explored and answered: (1) can computer software be trained to identify the gender origin of a subject according to one ultrasound image of their skeletal muscle? (2) If yes, what can we learn from the computer software classification by comprehending the core clues, or features, used?

The subjects in the experimental study included 107 healthy young volunteers (55 male and 52 female, age 21.0 ± 1.9 years, height 1.67 ± 0.71 m, weight 58.8 ± 10.6 kg, and body mass index (BMI) 21.0 ± 2.9).

For the dataset including 1498 ultrasound images of skeletal muscle, it is demonstrated that a computer application can be trained to classify muscle images from different genders using a CNN. The model achieved a high accuracy of 96.7% on the training dataset and 95.2% on the test dataset. The differences in image content between skeletal muscle of different gender groups were confirmed under ultrasound in the experiments. Furthermore, to determine the accuracy of neural network classification, a saliency map visualization tool has been employed. By overlapping the saliency maps (as shown in Fig. 7.2), it was observed that for males the features extracted in the middle and upper parts of the skeletal muscle images had a crucial influence on the classification results, while for images from female subjects, the saliency regions were more diverse (i.e., not as concentrated) and presented a more

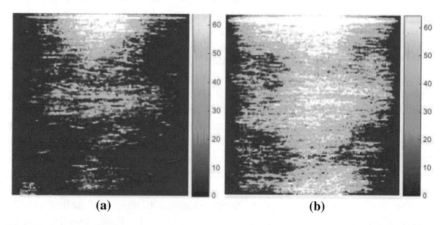

(a)   **(b)**

Fig. 7.2 Overlapped saliency maps for 16 single saliency maps. **a** Male and **b** Female Reprinted from A. Jx, A. Dx, B. Qw, and A. Yz, "Automatic classification of male and female skeletal muscles using ultrasound imaging—ScienceDirect," Biomedical Signal Processing and Control, vol. 57, 2020, with permission from Elsevier

complex analysis problem in comparison to those from male subjects. Whether such difference is systematic or not, in terms of how concentrated the saliency features are, has not been explored in the report.

Another preliminarily finding was about the effects of various occlusion designs on the classification accuracy rate. As the highlighted areas in the saliency maps appeared mostly in the middle and upper parts of the male muscle images, it was observed that the corresponding first occlusion method (with the lower part and the left and right sides masked out) improves the male classification accuracy to 100.0%. On the contrary for female muscle images, by the same occlusion design, the classification accuracy dropped sharply to 49.3%. Supposing that such a drop was due to the occlusion of the lower parts, a second occlusion design was presented for further verification in which only the left and right sides are occluded. In comparison with the first occlusion design, the second one led to a greatly improved classification accuracy of the female images, even slightly higher than the accuracy of classifying images without occlusion. These occlusion experiments have explored and extended our knowledge of the differences of images from different gender groups.

Now that the attempt to separate ultrasound images of skeletal muscle from two gender groups has achieved some success (the proposed method reached a classification accuracy of 95.2%), the natural next task was to determine if there were differences in ultrasound images of muscles under different physical/pathological conditions and, if so, can artificial intelligence classify them as well. This is going to be one of the most exciting and challenging topics in the field of musculoskeletal ultrasound.

7.3 Classification of Healthy Individuals with Different Exercise Levels Using Machine Learning

In 2020, Sun et al. demonstrated the feasibility of using ultrasound imaging and machine learning techniques to classify subjects with different exercise levels, to see if ultrasound images obtained from an individual have sufficient information reflecting the micro-structure changes induced by exercise [19]. Based on the same dataset from Sect. 7.2, a multiple-images-feature-selection (MIFS) framework was proposed to resolve this challenging classification problem by taking information from multiple images into consideration (shown in Fig. 7.3).

The MIFS framework mainly includes two parts: feature extraction and feature selection. Firstly, 54 image features (details are shown below) were extracted from the regions of interest (ROIs) of multiple images and combined with extra features (EF) to establish the total feature set. EF included not only physiological features but also features obtained through prior knowledge, which made the entire framework more scalable. In this study, EF consisted of gender and BMI. Through feature selection, important features were selected from the complete feature set and built into the

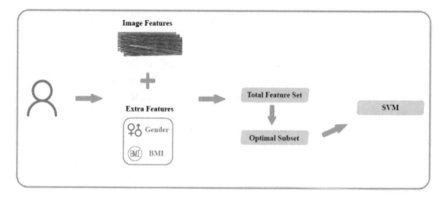

Fig. 7.3 MIFS workflow. Extra features contain physiological features and features obtained through prior knowledge. Reprinted from S. Sun, W. Xue, and Y. Zhou, "Classification of young healthy individuals with different exercise levels based on multiple musculoskeletal ultrasound images," Biomedical Signal Processing and Control, vol. 62, p. 102,093, 2020, with permission from Elsevier

optimal feature set. Ultimately, the SVM classifier was trained to take advantage of the features of the optimal feature set.

In MIFS, image features are extracted from multiple musculoskeletal ultrasound images (MUI), including muscle morphological features, image frequency analysis features, and image texture features:

- Muscle morphological features

  Sonography has been widely used in the study of skeletal muscle since its invention. Traditionally, muscle features such as muscle thickness, muscle fiber length, and pennation angle can be extracted from MUI. Muscle thickness (MT) is related to muscle mass, and usually defined as the distance between the midpoints of the upper and lower fascia. In this experiment, MT was calculated automatically through computer recognition of fascia, and a double check was performed manually to ensure the accuracy and validity of the data.

- Image frequency analysis features

  In recent years, researchers have also extracted mean frequency analysis features of the ROI (MFAF) and texture analysis features of the ROI (TAF) to analyze the texture information of muscle images. Since both MFAF and TAF are calculated from the ROI of the image in MUI, the appropriate ROI needs to be selected before feature extraction. For each image, a ROI was selected for analysis. The ROI was automatically chosen between the superior fascia and the central tendon by the computer program in MATLAB and was sized 128 × 64 pixels empirically. Finally, the location of ROI was verified and fine-tuned by two experienced sonographers. MFAF is an image frequency-domain analysis method, which does not fluctuate considerably with different scanning conditions and is associated with muscle quality. Nishihara et al. [20] suggested that the MFAF value for elderly individuals is relatively larger than those of youngsters, which implies that it may be an effective parameter for describing skeletal muscle structural differences. MFAF is a statistic of the power density spectrum as shown in Eq. 7.1.

$$MFAF = \frac{\sum_{i=0}^{m} I_i f_t}{\sum_{i=0}^{m} I_i} \tag{7.1}$$

where I, n, and f represent the power, length, and frequency of the power density spectrum, respectively.

- Image texture features

TAF refers to the texture analysis of the ROI of an ultrasound image, primarily containing first-order statistic (FOS) features as well as high-order texture features (e.g., Haralick features, Galloway features). Previous studies have found that FOS features can quantitatively and robustly describe the skeletal muscle ultrasound echo intensity. This study mainly utilized FOS features, Haralick features, Galloway features and Local Binary Pattern (LBP) features in this experiment.

(a) FOS features that are derived from the Gray-Level Histogram are the most common and prevalent image texture features. Seven features are used, including integrated optical density, mean, standard deviation, variance, skewness, kurtosis, and energy. These describe the ultrasound intensity information of muscle tissue.

(b) Haralick features are calculated from the Gray Level Co-Occurrence Matrix (GLCM). GLCM is a common way to describe image textures by studying the spatial correlation properties of grayscale. Typical Haralick features include contrast, correlation, energy, entropy, homogeneity as well as symmetry. Commonly, each Haralick feature contains four directions (0°, 45°, 90°, and 135°), so a total of 24 Haralick features are used for investigation. Details about the GLCM calculation can be found in Haralick [21].

(c) Galloway features are based on the Gray-Level Run-Length Matrix (GLRLM) from which several statistical measures can be obtained. GLRLM indicates the regularity of texture change in an image, whose dimensions are determined by the gray level of the image and the image size [22]. Some appropriate features describing image texture changes can be extracted from GLRLM, including Short Run Emphasis (SRE), Long Run Emphasis (LRE), Gray Level Non-Uniformity (GLNU), Run Length Non-Uniformity (RLNU), and Run Percentage (RP). Similar to Haralick features, each Galloway feature also contained four directions in this realization, and 20 Galloway features per ROI were obtained.

(d) The LBP features are obtained by comparing the center pixel in a local region of an image with its neighborhood. The LBP features mainly describe an image's local texture features and have significant advantages such as rotation invariance and gray invariance. Two LBP features, including energy and entropy, were chosen for analysis (Table 7.1).

Table 7.1 Mathematical description of the LBP features

LBP feature [1]	Description
Energy (LBP_{energy})	$LBP_{energy} = \sum_i f_i^2$
Entropy ($LBP_{entropy}$)	$LBP_{entropy} = \sum_i f_i^2 log_2(f_i)$

Annotation f_i, the relative frequency of the ith block in the local area of the image

Overall, 756 image features (54 features per ROI of the image and 14 images per subject) were obtained together with two EF for each of the subjects. To determine the optimal feature set and avoid overfitting, it is essential to reduce the number of training features via feature selection. In this experiment, Recursive Feature Elimination (RFE) as the feature selection method was adopted. The stability of the RFE algorithm depends on the classifier it uses. The linear-SVM with the L2-penalty was used, which made the method more robust. To improve the generalization of the selected features, the training framework inspired by Iizuka was adopted in this study [23].

Since features of different scales have a great influence on the performance of the SVM classifier, all features needed to be standardized before feature selection. Consequently, the values of the features were scaled to zero mean and one variance. Standardization and feature selection were performed by the machine learning package, Scikit-learn 0.20.0 (scikit-learn.org).

For classification, a typical machine learning model, support vector machine (SVM), was trained. SVM seeks the maximal bandwidth to separate the data, which is a quadratic programming problem. It makes some data that are inseparable in linear space separable in other dimensions through the kernel method. The algorithm itself is inherently robust, and regularization is introduced to make it more stable. Although many complex kernel methods (such as Gaussian kernel) have more powerful performance, linear kernel is used in this experiment because of its interpretability as well as simplicity. Actually, there was no significant difference in the performance of the Gaussian kernel and the linear kernel for SVM as found in this experiment.

Datasets are usually divided into training set, validation set, and test set when the amount of data is sufficient, and the final performance results come from the test set. However, the amount of data in this experiment was relatively small, which rendered the single test set mentioned above unable to reliably report the performance indicators of the classifier. Although this disadvantage can be solved by ten-fold cross-validation it may result in a hyperparameter-optimized dataset containing the data of the test set, making the evaluation a little optimistic. Therefore, nested cross-validation to tune and test the model, was adopted which can reduce the risk of data leakage and make the performance evaluation approximate to the real value. In addition, in order to calculate the AUC (area under the curve of ROC), the leave-one-out (LOO) protocol was not used. Actually, if the LOO protocol was used for evaluation in this experiment, most of the performance could increase by about 1%-2%. In order to more accurately analyze the performance of the classifier, the ROC (receiver operating characteristic) curves of all classifiers were calculated. On the

Table 7.2 Top 10 features in the optimal feature set

MFAF (action #4)*
Haralick correlation ($\theta = 90°$, action #1)
MFAF (action #3)*
RP($\theta = 45°$, action #7)
GLNU ($\theta = 135°$, action #4)*
Haralick correlation ($\theta = 135°$, action #7)
Haralick energy ($\theta = 45°$, action #1)
RLNU ($\theta = 0°$, action #1)
RLNU ($\theta = 95°$, action #7)
Haralick energy ($\theta = 0°$, action #4)**

Abbreviations MFAF, mean frequency analysis of the regions of interest; GLNU, gray-level non-uniformity; RLNU, run length non-uniformity; RP, run percentage; *, p value < 0.05; **, p value < 0.01

basis of that, the AUC was used to describe and represent the ROC curve for each classifier. AUC, the area enclosed by the ROC curve and the horizontal axis, was a robust indicator [24, 25].

In the linear-SVM model that was trained with the optimal feature set, the top 10 important features for the task are shown in Table 7.2. The most important features are the MFAF ones, the Haralick ones (correlation and energy), and the Galloway ones (gray-level non-uniformity, run length non-uniformity, run percentage, short run emphasis) while the FOS feature is not included. Besides, the correlation coefficient between features had a maximum value of 0.42, which was less redundant in the optimal feature set. Because the top 10 features all came from the normal distribution, the Student's t-test on regular exercisers and irregular exercisers in the top 10 features were conducted. Four features with significant differences are shown in Table 7.2 (indicated by asterisks). Two of them with p values less than 0.01 were MFAF (action #4) and Haralick energy ($\theta = 0°$, action #4). Compared to irregular exercisers except for Haralick energy ($\theta = 0°$, action #4) and GLNU ($\theta = 135°$, action #4), the two MFAFs of regular exercisers were smaller.

The amount of data is an important factor affecting the performance of the classifier. The stability of the classifier can be easily influenced by a small sample size. In this case, the performance of the classifier was improved by increasing the data size. When the classifier performance remains stable with an increase of data, the amount of data is appropriate. This procedure can be depicted by a learning curve. On a typical learning curve, the horizontal axis represents the number of samples in the training set while the vertical axis represents the error rate of the classifier in the test set. The logistic regression, SVM and neural network are trained with random training sets of different sizes and evaluated on the rest of the data that were not included in the training subsets. This step was repeated ten times and the average error rates were plotted. Figure 7.4 shows the learning curves of the linear-SVM model trained with the optimal feature set. As expected, the error rate of the trained

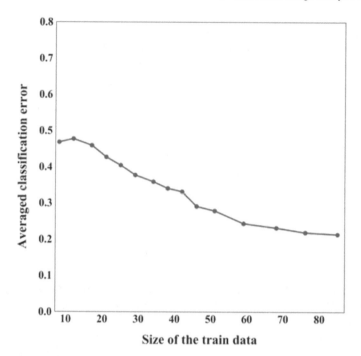

Fig. 7.4 The learning curves. Each point in each learning curve is the averaged classification error of the leave-one-subject-out protocol

classifier was reasonable and decreased as the number of training samples increased. The learning curve fairly fluctuated at the beginning. When the average classification error rate no longer decreased significantly with an increase of the training samples, the learning curve became stable gradually. In Fig. 7.4, when the training data of the linear-SVM model are larger than 68 samples, the performance of classifiers changes slowly. Therefore, after this observation point, the performance of the classifier could not be significantly improved with more data. The average classification error of SVM was 0.23 after training with 68 samples.

In this study, the classification of exercise level was poorly distinguished, when only one skeletal muscle ultrasound image information was used. Even with the state-of-the-art method of representing the muscle texture structure, the final result was not satisfactory for classification. In order to improve the performance, more image information was added. Using the proposed MIFS framework, images from different parts or actions were merged together to establish an optimal feature set. The results show that an optimal feature set can help distinguish between different exercise levels.

In reality, most tasks are usually more complicated. Especially in the field of sports rehabilitation, there are many factors affecting the state of sport or athletic ability, such as psychology, environment, etc. Lots of these factors are uncontrollable and cannot be digitized. Fortunately, muscle structure and texture as factors related to the state of sport can be represented by different methods. However, only through the muscle texture information using a single static image related to a certain status of action, it is difficult to understand complex motion tasks. The recognition of exercise levels is a relatively complicated task. To solve this problem, apart from proposing a better feature extraction method, increasing the ultrasound image of different motion states is also a direction worthy of attention. Since motion is a dynamic process, by analyzing and combining features in different motion states, it is possible to construct a feature set with more specific expression. The MIFS framework is based on this idea. However, the biggest challenge of this method is feature selection. Although the local optimal solution obtained by the feature selection algorithm is usually an approximated global optimal solution, it usually reduces the feature set dimension to mitigate the risk of overfitting when the sample size is small. In this study, the left and right leg data were combined through MANOVA analysis. To simplify the experiment and explore the optimal feature set stability, the same feature was utilized to train the linear-SVM model on the left leg data and the right leg data.

It is worth mentioning that physical activity is usually quantified using the International Physical Activity Questionnaire (IPAQ). The results of the IPAQ can be reported in categories (low, moderate, or high activity levels) through the statistical analysis of subjects' total weekly physical activity levels. This result reflects the total physical activity level, but this tentative study mainly focused on a specific ankle-based exercise. In addition, the influence of other activities on muscles were assumed to be negligible since the frequency and intensity of daily physical activity of subjects are much higher than those of other types of exercise in this experiment. Therefore, a simplified criterion for regular exercise was proposed for the experiments. In this study, the standard for judging the exercise levels of regular exercisers adopted the moderate activity levels in the IPAQ and the intensity of exercise was limited by the RPE value (RPE ≥ 5), which made the experiment focus on exploring the relationship between ankle-based exercise and leg muscle structure.

Although the accuracy of the MIFS model reached 79.8%, the number of the TN for all three classifiers was higher than the number of the TP, which indicated that the classifier was easier to identify the irregular exercisers and was relatively insensitive to the regular exercisers. Regarding the results, there were the following three speculations: (1) Other types of exercise slightly interfered with the results of this experiment; (2) RPE was not objective enough to quantify exercise intensity, which may cause a certain deviation of the label; and (3) Since the effect of exercise on muscle structure is slow in the short term, it was also very subtly reflected in the ultrasound image. Nonetheless, the result still indicated that the exercise levels of different individuals can be distinguished by the analysis of muscle ultrasound images. In the future, the criteria can be improved to make the research more comprehensive and rigorous by using the IPAQ to investigate the exercise status of subjects, and using objective indicators such as heart rate to quantify exercise intensity, so as to evaluate the MIFS more generally.

Since high-order texture and frequency information are more sensitive to capturing such tiny changes in muscle structure, they have the potential for the broader application to problems that are both sensitive and specific to distinguishing subtle changes in skeletal muscle structure. Beyond judging the impact of regular exercise, these features may also be used to assess the basic motor ability of muscles in the future. Moreover, the quantitative evaluation of the exercise capacity of muscles through ultrasound imaging needs to be explored to identify the indicators which can quantify the general motor ability.

Finally, this work has the following two limitations. On one hand, the experimental objects were not diversified, which may affect the generalization of the model. In future research, datasets can be expanded to improve data diversity and model robustness. On the other hand, exercise is a systemic activity, but current experimental data is limited to ankle-based exercise. Thus, for different movements, ultrasound image data of different parts should be collected to implement comprehensive research in future study.

To summarize, there are two contributions for the classification of exercise levels in this preliminary study. One is that a multiple image-based MIFS framework is proposed to solve the difficult task which is difficult to handle using only a single image. The other is that a classifier is proposed, based on ultrasound images, for subjects with different exercise levels, and an optimal feature set is obtained. Although the exercise in this study was limited to ankle-based exercise, the framework and research methods proposed are expected to be re-applicable for the study of various exercise levels, by coupling the slight muscle changes induced by exercise with the ultrasound images of muscles. In the future, the optimal feature set can be used to establish the connection between ultrasound and muscle motor ability, and even hopefully push forward the quantification of muscle exercise effects.

7.4   Ultrasound Assisted with Machine Learning in LBP Study

Lower back pain (LBP) is one of the most common musculoskeletal diseases with 90% of LBP being nonspecific lower back pain (NSLBP) without specific pathology [26, 27]. At present, a visual analog scale (VAS) is widely used to assess LBP. The higher the VAS score, the higher the degree of pain. We recently proposed a machine learning model to fit the qualitative VAS index through the ultrasound image of the lumbar multifidus muscle, so as to reduce the influence of subjective factors and provide an important auxiliary assessment method for making treatment plans or accurate evaluation of the treatment effects on LBP.

Data were collected from 13 subjects (five females, eight males, BMI: 23.3 $\pm$ 2.54) recruited from Shenyang North Hospital, China). All subjects were screened before participating in this study, and the inclusion criteria were as follows: Patients with moderate or above intensity of LBP (VAS $\geq$ 2) during rest or daily activities. The exclusion criteria were spine or lower limb fracture history, spine surgery history, or spine deformity history. Ethical approval was granted by the authority of the corresponding institute, and the written informed consent of all subjects was obtained.

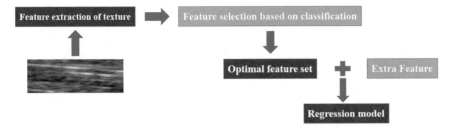

Fig. 7.5 Reg-SIFS framework flow chart

Two of the subjects dropped out of the trial due to force majeure, and finally a total of 11 subjects completed the test. In addition, each subject was required to collect two sets of test data, respectively, at the first visit and after 7 days of conventional physical therapy. Therefore, there were 22 valid trials in this experiment (the data of the same subject before and after treatment is counted as two trials). For each subject, a total of four ultrasound images from different locations were collected, and the pain area was marked down. Assuming that the pain areas were independent of each other, the four ultrasound images of the lumbar multifidus collected by each subject can be regarded as four independent samples. The pain area corresponds to its VAS score, and the default VAS of other areas is 0. Finally, the number of data samples counted was 88.

In order to solve the problems of small sample size and poor stability of VAS score, we proposed a regression-single-images-feature-selection (Reg-SIFS) framework. Compared with traditional single images feature selection (SIFS), Reg-SIFS improves the performance of the model in small datasets by introducing prior knowledge. The framework flow chart is shown in Fig. 7.5

In addition, the Reg-SIFS framework divides VAS into two intervals by using prior knowledge in feature selection. Among them, VAS = 4 is the threshold (prior knowledge). Cases lower than this are classified into one class, and the others are classified as another class, so that the problem is transformed into a classification task in feature selection. Therefore, compared with traditional SIFS, the reg SIFS framework has three advantages:

(1) The error of the original VAS score in the feature selection stage is reduced because the variations in VAS score are relatively scaled down [28].
(2) The model precision is further improved by introducing prior experience. VAS = 4 is regarded as the standard provided by the VAS system itself, and the difference of pain feeling between patients, with VAS larger or smaller than the threshold value, is found to be distinct in practice.
(3) There are some differences of labeling between the feature selection and model training stages, which improves the possibility of mining new information.

In addition, inspired by the Attention U-Net network framework proposed by Oktay et al. (Fig. 7.6) [29] , we made the following three main improvements:

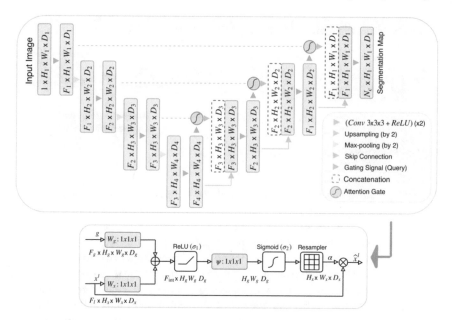

Fig. 7.6 Attention U-Net network [29]. Reprinted from O. Oktay et al., "Attention u-net: Learning where to look for the pancreas," arXiv preprint arXiv:1804.03999, 2018

(1) The padding was set so that the image size did not change after each convolution, thus simplifying the training optimization process.

(2) By using 2 * 2 convolution kernel instead of pooling operation, the original information was kept as much as possible, the loss of detailed information was reduced, and the convergence speed of the network was accelerated.

(3) The batch normalization mechanism was introduced to optimize the parameter adjustment process, improve the generalization of the network, and use a larger learning rate to further improve the network training speed.

Then a fully automatic method was proposed to calculate the morphological parameters of ultrasonic images, and the ROI of the lumbar multiform muscle was automatically cut. The segmentation results were measured by the Dice coefficient. Finally, in the test stage, the Dice coefficient of the network was 0.84. The segmentation result is shown in Fig. 7.7.

In Reg-SIFS, in addition to prior knowledge features, also known as extra features (EF), there are three additional kinds of image features: muscle shape features, image average frequency features, and image texture features. Totally 55 image features and one EF feature (muscle stiffness) are obtained. Meanwhile, in order to determine the best feature collection and avoid over fitting, we adopted recursive feature elimination (RFE) [30] as the feature selection method.

With different frameworks, three training feature sets are obtained: full set feature set, SIFS feature set, and Reg-SIFS feature set. Among them, the full feature set includes all features without feature selection; the SIFS feature set includes the best

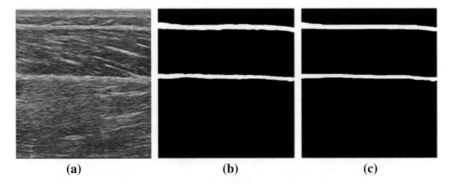

(a)   **(b)**   **(c)**

Fig. 7.7 **a** Raw ultrasound image of muscle. **b** Image annotated by experts. **c** Segmentation result by the improved attention U-Net

feature set obtained by the SIFS framework; the Reg-SIFS feature set includes the best feature set obtained by using the Reg-SIFS framework. These three feature sets are used to train Support Vector Regression (SVR), Bayesian, Random Forest, Adaptive boosting (AdaBoost), Gradient Boosting Decision Tree (GBDT), and bagging regression models. The data volume ratio of the training set and test set is 7:3. All models were evaluated on the test set, and R^2 and mean square error (MSE) were used as evaluation indexes.

As shown in Fig. 7.8, the R^2 value of the model trained by the Reg-SIFS feature set is the largest in all three feature sets, and the R^2 value of AdaBoost model reached 0.99, indicating that the model has a very strong correlation. The second is the MIFS feature set. Although the R^2 of the MIFS feature set is smaller than that of the Reg-SIFS feature set, its performance is better than that of the model without feature selection. The R^2 value of AdaBoost model is generally higher, which exceeds 0.5. The R^2 values of Random Forest, GBDT, and bagging were all over 0.5 except for the full feature set.

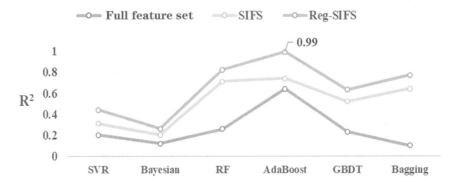

Fig. 7.8 The R^2 in three features sets of six models

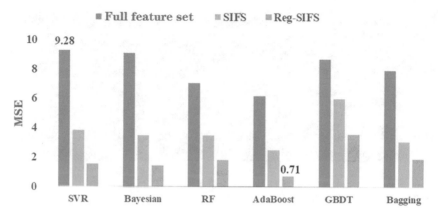

Fig. 7.9 The MSE in three feature sets of six models

Compared with the goodness of fit of the R^2 evaluation model, MSE can more directly reflect the prediction accuracy of regression models. Figure 7.9 shows the MSE of the six models in different features. The AdaBoost model based on the Reg-SIFS framework also achieved the optimal MSE-0.71, so it can be considered as the optimal model for this task. In addition, it can be seen that the performance of the model trained by SIFS is generally better, and the improvement of Reg-SIFS on the model is also huge.

With an application example of LBP investigation, we have initially demonstrated the application progress of SMG combined with AI in the intelligent assessment of muscle disease states. This field is developing rapidly. AI and other engineering technologies, combined with ultrasound, will surely promote the objective quantification of muscle research.

7.5   An Integrated Multi-model Sensing System for Muscle Study

In the current era of big data and big health, artificial intelligence technology can truly have sufficient generalization and practicality only after it is scaled up on data. And inexpensive, easy-to-operate, and integrated medical information sensing technology may be a must for the scale of data. Motion capture and analyzing systems are essential for understanding locomotion. However, the existing devices are too cumbersome and can be used indoors only. A newly-developed wearable motion capture and measurement system with multiple sensors and ultrasound imaging was introduced in a recent study [24].

In ten healthy participants, the changes in muscle area and activity of the gastrocnemius, and plantarflexion and dorsiflexion of the right leg during walking were

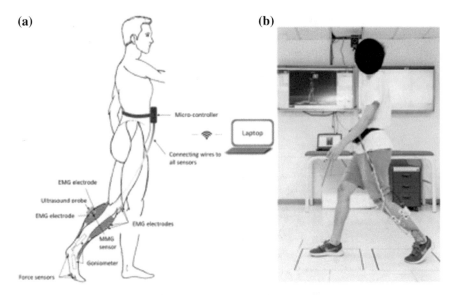

Fig. 7.10 a An illustration of the wearable mobile sensing system with real-time ultrasound imaging and location of the ultrasound probe, electromyography (EMG) electrode, mechanomyography (MMG) electrode, force sensors, and goniometer at shank and foot. **b** A demonstration of a subject wearing a wearable mobile sensing system. Reprinted with permission from C. Z.-H. Ma, Y. T. Ling, Q. T. K. Shea, L.-K. Wang, X.-Y. Wang, and Y.-P. Zheng, "Towards wearable comprehensive capture and analysis of skeletal muscle activity during human locomotion," Sensors, vol. 19, no. 1, p. 195, 2019. https://doi.org/10.3390/s19010195

evaluated by the developed system and the Vicon system. The existence of significant changes in a gait cycle, comparison of the ankle kinetic data captured by the developed system and the Vicon system, and test–retest reliability (evaluated by the intraclass correlation coefficient, ICC) in each channel's data captured by the developed system (Fig. 7.10) were examined. The flowchart is shown in Fig. 7.11.

The data were analyzed using Matlab (version 2016b, The MathWorks Inc., Natick, MA, USA). MMG and EMG data were filtered using the 4th-order Butterworth band-pass filter of 5–50 Hz and 30–500 Hz, respectively. The signals were then rectified and filtered with a moving-average filter of temporal window of 0.101 s. Baseline offsets of the recorded angles and forces were removed with subtractions of the steady-state baseline values. In the ultrasound images, muscle boundaries of the gastrocnemius muscle were indicated manually by drawing two lines in each frame by a trained practitioner. The muscle area was then calculated from the area between the two lines in each frame. The measured muscle area at heel strike was set as the baseline to calculate the changes in muscle area in a gait cycle. The step of interest was extracted manually from the time-series data with reference to the plantar forces. In calculation of stance and swing time, the end of stance phase is regarded as the time when the force detected falls below 1% of their peak values. The data was then resampled to 0–100% of gait cycle using piecewise cubic interpolation. EMG

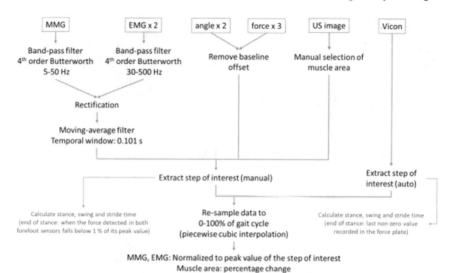

Fig. 7.11 Flow chart for data processing and analysis. [31] Reprinted with permission from C. Z.-H. Ma, Y. T. Ling, Q. T. K. Shea, L.-K. Wang, X.-Y. Wang, and Y.-P. Zheng, "Towards wearable comprehensive capture and analysis of skeletal muscle activity during human locomotion," Sensors, vol. 19, no. 1, p. 195, 2019. https://doi.org/10.3390/s19010195

and MMG signals were normalized to their peak values during walking. Percentage change of muscle area with respect to the beginning of a gait cycle was calculated. The forces measured at the first metatarsal head, the second metatarsal head, and the heel were summed to calculate the plantar force of a full foot, and normalized by dividing by each subject's body mass. Vicon data were extracted for the step of interest with reference to the foot progress angle in the sagittal plane. In the calculation of stance and swing time, the end of stance phase is regarded as the time when the force detected by the force plate returns to zero. The step was then resampled to 0–100% of gait cycle using piecewise cubic interpolation. Therefore, the step data from two systems were synchronized by the defined gait cycle. The plantar force data captured by the force sensors in the wearable mobile sensing system were compared to the plantar force captured by the floor-mounted force plate in the Vicon system. The ankle joint angles (plantarflexion/dorsiflexion and inversion/eversion) captured by the goniometers in the wearable mobile sensing system were compared to the corresponding joint angle captured by the Vicon system.

The measured ankle activities in a gait cycle measured by the newly developed mobile SMG system and the Vicon system of 10 healthy subjects are illustrated in Fig. 7.12. The trend of ankle plantar-/dorsi-flexion and in-/e-version measured by the mobile SMG device was similar to that measured by the Vicon system. Significantly high correlation between the goniometer of the introduced mobile SMG system and the Vicon system was found in peak plantarflexion ($R = 0.703$, $p = 0.023$) and peak dorsiflexion ($R = 0.707$, $p = 0.022$) during the swing phase, as well as the peak eversion ($R = 0.638$, $p = 0.047$) during the initial stance phase.

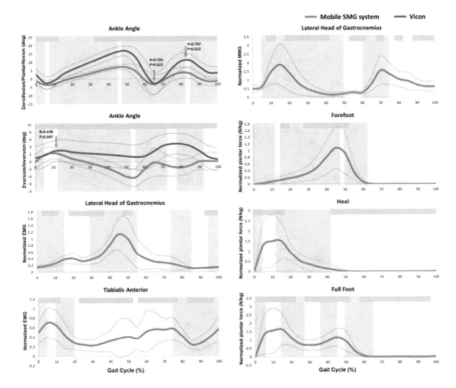

Fig. 7.12 Ankle activities in a gait cycle measured by Vicon and the newly developed mobile SMG system [24]. The bolded blue line illustrates the averaged data measured by the Vicon system; the thin dashed line illustrates the standard deviation (SD) of each corresponding bold line. There was a significantly high correlation in intraclass correlation coefficient (ICC) among three trials; the bolded orange line illustrates the averaged data measured by the mobile SMG system; the bolded blue line illustrates the averaged data measured by the Vicon system; the thin dashed line illustrates the standard deviation (SD) of each corresponding bold line. Reprinted with permission from C. Z.-H. Ma, Y. T. Ling, Q. T. K. Shea, L.-K. Wang, X.-Y. Wang, and Y.-P. Zheng, "Towards wearable comprehensive capture and analysis of skeletal muscle activity during human locomotion," Sensors, vol. 19, no. 1, p. 195, 2019. https://doi.org/10.3390/s19010195

The novel wearable mobile sensing system can also show that muscle area remained below the baseline muscle area during the whole gait cycle (Fig. 7.13). It firstly reduced after heel-strike, and reached a trough (6.28% ± 3.85%, p = 0.001) in mid-stance, followed by an increase until the terminal-stance phase. This muscle area reduced again from the pre-swing phase and reached a minimal muscle area (13.6% ± 3.8%, p = 0.002) during the initial swing phase, followed by an increase until the terminal-swing phase in subjects.

The minimal muscle area of a gait cycle was found in the initial swing phase, which was 142.4% less than the trough in the mid-stance phase (p < 0.001).

Moderate to good test–retest reliability of various measurement channels of the mobile SMG system and strong correlation between this system with the Vicon

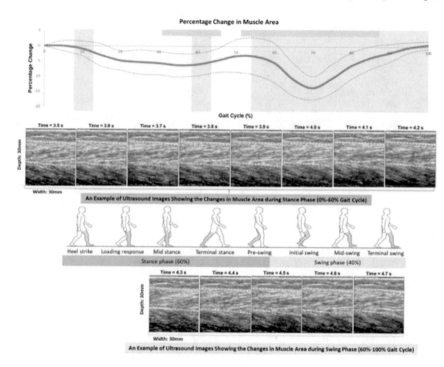

Fig. 7.13 Percentage changes in muscle area in a gait cycle measured by the newly developed mobile SMG system. [24] Reprinted with permission from C. Z.-H. Ma, Y. T. Ling, Q. T. K. Shea, L.-K. Wang, X.-Y. Wang, and Y.-P. Zheng, "Towards wearable comprehensive capture and analysis of skeletal muscle activity during human locomotion," Sensors, vol. 19, no. 1, p. 195, 2019. https://doi.org/10.3390/s19010195

system regarding the ankle plantar-/dorsi-flexion angle measurement were found. The averaged ICC values of EMG, goniometer, and force sensors were found to be 0.75 or above, suggesting good test–retest reliability in these measurement channels. Meanwhile, moderate test–retest reliability was found in MMG and ultrasound imaging ($0.5 < ICC < 0.75$, $p < 0.05$). The ankle plantar-/dorsi-flexion angle measured by the mobile SMG device was comparable to that of the Vicon system, and strong correlation in peak dorsiflexion and peak plantarflexion angle during the swing phase was found between the two systems ($R > 0.5$, $p < 0.05$). The ankle inversion/eversion angle measured by the mobile SMG device was not quite comparable to that of the Vicon system. One possible reason could be the different axis of rotation in angle measurement, i.e., the Vicon system estimated the angle from the built-in model, while the goniometer measured the angle from the medial side of the ankle joints.

The novel mobile sensing system developed for gait analysis can be a powerful tool in muscle studies. In addition to supporting the moderate to good test–retest reliability of this system, the findings of this study also revealed some features of muscle activity, especially the gastrocnemius, during walking under comprehensive data capture and analysis.

Such a system with real-time ultrasound imaging would enable the comprehensive investigation of real-time muscle behavior and adaption during different activities in vivo in humans. The success of this project will pave the way for real-time in-vivo analysis of muscle activity in both indoor and outdoor environments, which offers tremendous potential health benefits. This device can be widely applied to evaluate human posture and motion, as well as the treatment outcomes of patients in the future. It is a systematic extension of the team's previous research, which utilizes its unique expertise in ultrasound imaging technology, as well as posture and gait analysis in various populations. This mobile system and its successors will greatly reduce the threshold for obtaining multi-modal muscle big data and provide an inexpensive entry path for subsequent data-driven AI muscle research.

References

1. Cunningham, R., Sánchez, M.B., May, G., Loram, I.: Estimating full regional skeletal muscle fibre orientation from b-mode ultrasound images using convolutional, residual, and deconvolutional neural networks. J. Imaging **4**(2), 29 (2018). [Online]. Available https://www.mdpi.com/2313-433X/4/2/29
2. Jx, A., Dx, A., Qw, B., Yz, A.: Automatic classification of male and female skeletal muscles using ultrasound imaging—sciencedirect. Biomed. Signal Process. Control **57** (2019)
3. Janssen, I., Heymsfield, S.B., Wang, Z.M., Ross, R.: Skeletal muscle mass and distribution in 468 men and women aged 18–88 yr. J. Appl. Physiol. **89**(1), 81 (2000)
4. Stevens, J.A.: Gender differences for non-fatal unintentional fall related injuries among older adults. Injury Prevent. **11**(2), 115–119 (2005)
5. Ling, S., Zhou, Y., Chen, Y., Zhao, Y.-Q., Wang, L., Zheng, Y.-P.: Automatic tracking of aponeuroses and estimation of muscle thickness in ultrasonography: a feasibility study. IEEE J. Biomed. Health Inform. **17**(6), 1031–1038 (2013)
6. Callahan, D.M., et al.: Chronic disuse and skeletal muscle structure in older adults: sex-specific differences and relationships to contractile function. Am. J. Physiol. Cell Physiol. **308**(11), C932–C943 (2015)
7. Zhou, Y., Zheng, Y.-P.: Estimation of muscle fiber orientation in ultrasound images using revoting hough transform (RVHT). Ultrasound Med. Biol. **34**(9), 1474–1481 (2008)
8. Li, J., Zhou, Y., Lu, Y., Zhou, G., Wang, L., Zheng, Y.-P.: The sensitive and efficient detection of quadriceps muscle thickness changes in cross-sectional plane using ultrasonography: a feasibility investigation. IEEE J. Biomed. Health Inform. **18**(2), 628–635 (2013)
9. Chen, X., Zheng, Y.-P., Guo, J.-Y., Zhu, Z., Chan, S.-C., Zhang, Z.: Sonomyographic responses during voluntary isometric ramp contraction of the human rectus femoris muscle. Eur. J. Appl. Physiol. **112**(7), 2603–2614 (2012)
10. Zhou, Y. et al.: Quantitative comparison of muscle thickness between young male and female subjects using ultrasonography. In: 2016 IEEE-EMBS International Conference on Biomedical and Health Informatics (BHI), pp. 284–287. IEEE, (2016)
11. Zhou, Y., Yang, X., Yang, W., Shi, W., Cui, Y., Chen, X.: Recent progress in automatic processing of skeletal muscle morphology using ultrasound: a brief review. Current Med. Imaging **14**(2), 179–185 (2018)
12. Wang, C.-Z., Guo, J.-Y., Li, T.-J., Zhou, Y., Shi, W. , Zheng, Y.-P.: Age and sex effects on the active stiffness of vastus intermedius under isometric contraction. BioMed Res. Int. 2017 (2017)
13. Puthucheary, Z.A., et al.: Qualitative ultrasound in acute critical illness muscle wasting. Crit. Care Med. **43**(8), 1603–1611 (2015)

14. Paris, M., Mourtzakis, M.: Assessment of skeletal muscle mass in critically ill patients: consid-erations for the utility of computed tomography imaging and ultrasonography. Curr. Opin. Clin. Nutr. Metab. Care **19**(2), 125–130 (2016)

15. Parry, S.M. et al.: Ultrasonography in the intensive care setting can be used to detect changes in the quality and quantity of muscle and is related to muscle strength and function. J. Critical Care **30**(5), 1151. e9–1151. e14 (2015)

16. Litjens, G., et al.: A survey on deep learning in medical image analysis. Med. Image Anal. **42**, 60–88 (2017)

17. Shen, D., Wu, G., Suk, H.-I.: Deep learning in medical image analysis. Annu. Rev. Biomed. Eng. **19**, 221–248 (2017)

18. He, K., Zhang, X., Ren, S., Sun, J.: Delving deep into rectifiers: Surpassing human-level performance on imagenet classification. In: Proceedings of the IEEE International Conference on Computer Vision, pp. 1026–1034. (2015)

19. Sun, S., Xue, W., Zhou, Y.: Classification of young healthy individuals with different exercise levels based on multiple musculoskeletal ultrasound images. Biomed. Signal Process. Control **62**, 102093 (2020)

20. Nishihara, K., Kawai, H., Hayashi, H., Naruse, H., Hoshi, F.: Frequency analysis of ultrasonic echo intensities of the skeletal muscle in elderly and young individuals. Clinical Intervent. Aging **9**(default), 1471–1478 (2014)

21. Haralick, R.M.: Textural features for image classification. IEEE Trans. Syst. Man Cybernet. SMC **3** (1973)

22. Galloway, M.: Texture analysis using gray level run lengths. Comput. Graphics Image Process. **4**(2), 172–179 (1975)

23. Iizuka, N., et al.: Oligonucleotide microarray for prediction of early intrahepatic recurrence of hepatocellular carcinoma after curative resection. The lancet **361**(9361), 923–929 (2003)

24. Ma, C.Z.-H., Ling, Y.T., Shea, Q.T.K., Wang, L.-K., Wang, X.-Y., Zheng, Y.-P.: Towards wear-able comprehensive capture and analysis of skeletal muscle activity during human locomotion. Sensors **19**(1), 195 (2019)

25. Bradley, A.P.: The use of the area under the ROC curve in the evaluation of machine learning algorithms. Pattern Recogn. **30**(7), 1145–1159 (1997)

26. Hoy, D., et al.: Measuring the global burden of low back pain. Best Pract. Res. Clin. Rheumatol. **24**(2), 155–165 (2010)

27. Cheung, W.K., Cheung, J.P.Y., Lee, W.-N.: Role of ultrasound in low back pain: a review. Ultrasound Med. Biol. **46**(6), 1344–1358 (2020)

28. Langley, G., Sheppeard, H.: The visual analogue scale: its use in pain measurement. Rheumatol. Int. **5**(4), 145–148 (1985)

29. Oktay, O. et al.: Attention u-net: Learning where to look for the pancreas. (2018) *arXiv preprint* arXiv:1804.03999

30. Guyon, I., Weston, J., Barnhill, S., Vapnik, V.: Gene selection for cancer classification using support vector machines. Mach. Learn. **46**(1), 389–422 (2002)

31. Galloway, M.: Texture analysis using gray level run lengths. Comput. Gr. Image Process. **4**(2), 172–179 (1975)

Chapter 8
Future Perspectives of Sonomyography

Abstract A major advantage of muscle ultrasound is its capacity to capture muscle contraction with decent spatial and temporal resolutions. Ever since the first report of sonomyography, a number of related reports have been put forward by different research groups. Future perspectives of sonomyography, including the development of a multi-modal muscle research platform that integrates ultrasound, EMG, acceleration sensor signals, etc., are discussed in this chapter.

A major advantage of muscle ultrasound is its capacity to capture muscle contraction with decent spatial (it can assess deeper layers of muscles) and temporal resolution (up to over 100 frames per second, or even several thousands of frames per second in the case of sonomechanomyography (SMMG)). With the rapid development of computer vision techniques, especially the more recent prevalence of artificial intelligence techniques, applications of muscle ultrasound are to see vast progress, in terms of both more accurate and more comprehensive understanding and evaluation of skeletal muscle.

With the help of Web of Science, a plot of the number of reports till the year 2021 with the keywords of "skeletal muscle ultrasound image" can be visualized in Fig. 8.1.

Today, there are several parallel trends happening in the field of skeletal muscle studies with ultrasound. Firstly, computerized methods are gradually replacing the traditional manual method in image analysis, which is making possible video stream analysis to reveal muscle contraction dynamics, as well as big data applications involving many thousands of frames in muscle ultrasound imaging. Secondly, the dimension of information is expanding, not only from 1D ultrasound signals to 2D image and 3D ultrasound volume, but also by engaging other sensory systems, such as EMG, MMG, muscle force/torque, etc. Such an integrated data collection system helps to investigate skeletal muscle in a more comprehensive and penetrating way, and it is expected that the diagnosis or study based on such a system would be more precise and accurate. Thirdly, the borders of studies have also been expanded, from healthy subjects to subjects with pathological conditions, such as hemiplegia, muscle injuries, muscular dystrophy, etc. Methods to classify these subjects are being developed with the help of machine learning techniques, and new approaches to detect the stage of

© Springer Nature Singapore Pte Ltd. 2021
Y. Zhou and Y.-P. Zheng, *Sonomyography*, Series in BioEngineering,
https://doi.org/10.1007/978-981-16-7140-1_8

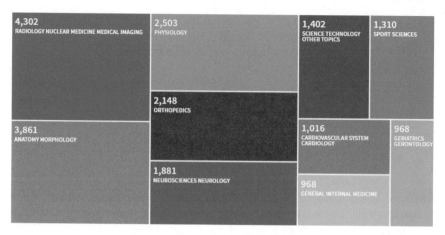

Fig. 8.1 Distribution of number of publications in various research areas related to "skeletal muscle ultrasound image" from the Web of Science database (2000–2021)

these conditions are expected now. Fourthly, the potentials of ultra-fast ultrasound imaging for the assessment of muscle contraction onset time have been recently demonstrated [1–3], and many further studies can be followed up along the direction of sonomechanomyography (SMMG), which detects the vibration of muscle during contraction using ultrasound imaging with a frame rate of several thousand frames/s. Last but not least, new techniques and findings in related areas such as biomechanics, metabolism, and the anatomy of the skeletal muscle system altogether are pushing the understanding of the system forwards, and then new applications in sports science and rehabilitation arise accordingly, which actively and significantly promote technique development.

The rapid progress of the field can also be partially reflected by the increasing number of publications and citations too, as shown in Fig. 8.2.

After more than a decade of development, today, with the increasingly close interaction of engineering technologies (especially computer vision and artificial intelligence) with clinical problems in skeletal muscle study, more and more successful applications are expected to be realized and deployed. Meanwhile, the development of ultrasound imaging technology is also booming, new imaging technology, with more customized probe designs (for example concave surface probe for better coupling with muscle surface), higher frame rate, higher quality, and more versatile packaging (such as palm-sized wireless ultrasound scanners) also provide more and more powerful supports for muscular ultrasound research and expanding its application domains. Last but not least, the development of a multi-modal muscle research platform that integrates ultrasound, EMG, acceleration sensor signals, etc., can also greatly promote the research of muscular ultrasound. Once such a platform becomes wearable, it will greatly enhance the approaches for functional muscle assessment for various purposes [4]. Starting with this book, we sincerely hope that colleagues from both engineering and medical sides can work together more to have better

Fig. 8.2 Distribution of publications per year (1995–2020) related to "skeletal muscle ultrasound image" from the Web of Science database

understanding, make more precise diagnoses and more personalized rehabilitation prescriptions with more objective and standardized protocols for skeletal muscle assessment.

References

1. Begovic, H., Zhou, G.Q., Li, T. J., Wang, Y., Zheng, Y.P.: Detection of the electromechanical delay and its components during voluntary isometric contraction of the quadriceps femoris muscle. Front. Physiol. **5**(UNSP 494), (2014)
2. Begovic, H., Zhou, G.Q., Schuster, S., Zheng, Y.P.: The neuromotor effects of transverse friction massage. Man. Ther. **26**, 70–76 (2016)
3. Ling, Y.T., Ma, C.Z., Shea, Q.T.K., Zheng, Y.P.: Sonomechanomyography (SMMG): Mapping of skeletal muscle motion onset during contraction using ultrafast ultrasound imaging and multiple motion sensors. Sensors. **20**(19), 1–13 (2020)
4. Ma, C.Z.-H., Ling, Y.T., Shea, Q.T.K., Wang, L.-K., Wang, X.-Y., Zheng, Y.-P.: Towards wearable comprehensive capture and analysis of skeletal muscle activity during human locomotion. Sensors **19**(1), 195 (2019)